W9-AUL-146

Grade **1**

KUMON MATH WORKBOOKS

Geometry & Measurement

Table of Contents

KUMON

Numbers up to 10

Level ★

Date   /   /

Name

Score   /100

1 Count the dots. Write the number in each box.

3 points per box

(1)

・ ・・ ・・・ ・・・・ ・・・・・

1   2   3   4   5

(2)

・・・・ ・・・ ・・ ・

4   3   2   1   Zero 0

(3)

6   7   8   9   10

(4)

10   9   8   7   6

 © Kumon Publishing Co., Ltd.

2 How many objects are in each picture? Write the number in each box below.

4 points per question

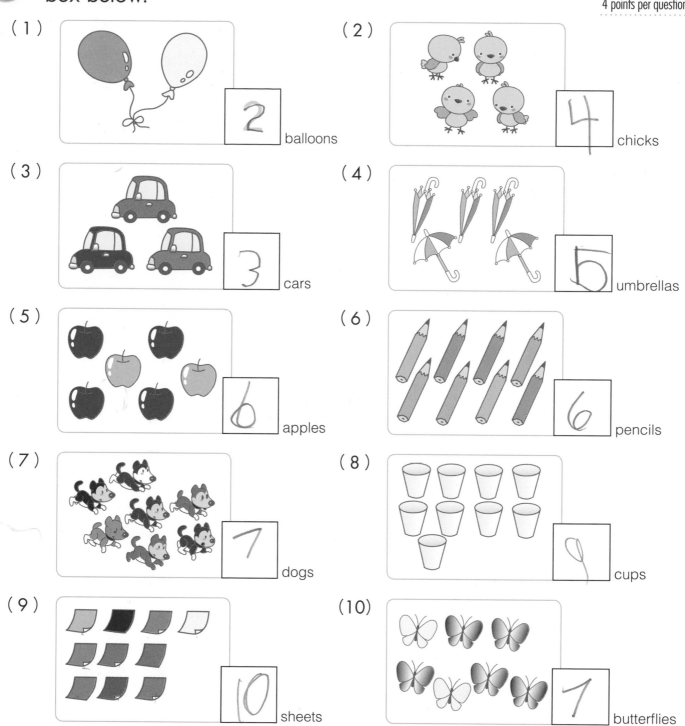

(1) $\boxed{2}$ balloons

(2) $\boxed{4}$ chicks

(3) $\boxed{3}$ cars

(4) $\boxed{5}$ umbrellas

(5) $\boxed{6}$ apples

(6) $\boxed{6}$ pencils

(7) $\boxed{7}$ dogs

(8) $\boxed{9}$ cups

(9) $\boxed{10}$ sheets

(10) $\boxed{7}$ butterflies

Counting is fun, right? Way to go!

© Kumon Publishing Co., Ltd.

Numbers up to 10

1 How many objects are in each picture? Write the number in each box below.

8 points per question

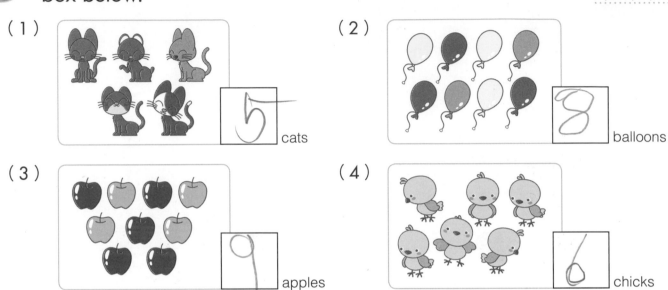

(1)   5 cats

(2)   8 balloons

(3)   9 apples

(4)   6 chicks

2 Color the correct amount of ○ to match each number in the box on the left.

8 points per question

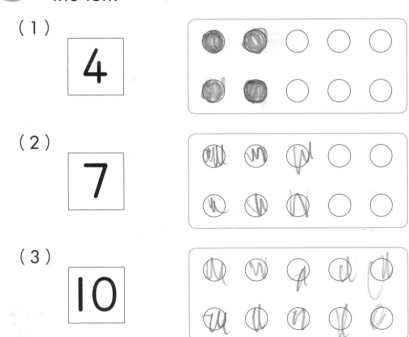

(1)   4

(2)   7

(3)   10

  © Kumon Publishing Co., Ltd.

3 Count the dots. Write the number in each box.

5 points per box

| 3 | 2 | 1 | 0 |

4 How many objects are in each picture? Write the number in each box below.

8 points per question

(1)

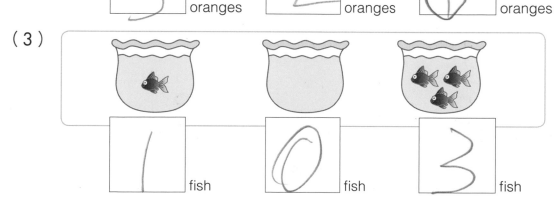

3 apples   2 apples   1 apple   0 apples

(2)

3 oranges   2 oranges   0 oranges

(3)

1 fish   0 fish   3 fish

Is this making you hungry?

© Kumon Publishing Co., Ltd.   5

3 Numbers up to 10

Date / /   Name

1   Write the missing number in each box.

6 points per question

(1) 1 2 3 4 5

(2) 6 7 8 9 10

(3) 10 9 8 7 6

(4) 4 3 2 1 0

2   Write the missing number in each box.

9 points per question

(1) 1 2 3 4 5 6 7 8 9 10

(2) 9 9 8 7 6 5 4 3 2 1

 © Kumon Publishing Co., Ltd.

3 Write the missing number in each box.

6 points per question

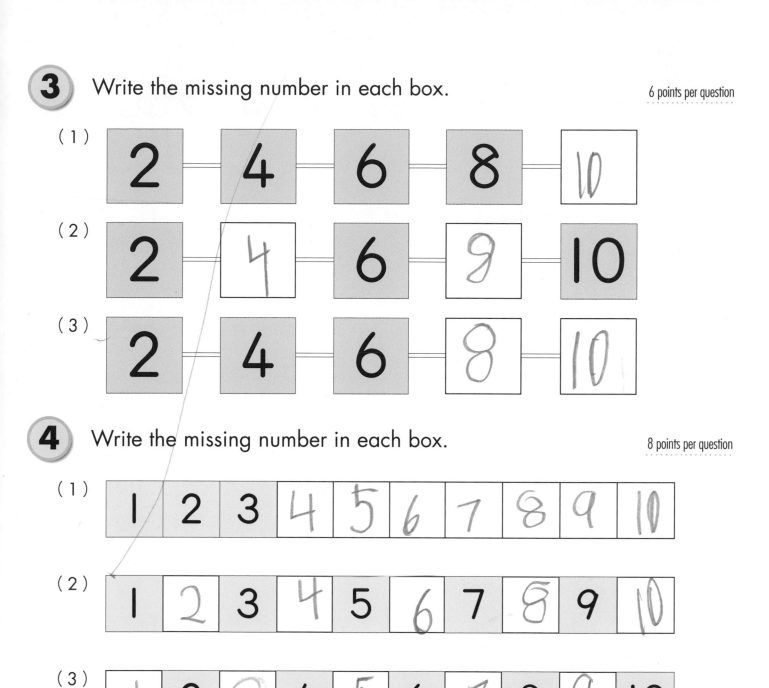

(1) 2　4　6　8　10

(2) 2　4　6　9　10

(3) 2　4　6　8　10

4 Write the missing number in each box.

8 points per question

(1) 1　2　3　4　5　6　7　8　9　10

(2) 1　2　3　4　5　6　7　8　9　10

(3) 1　2　3　4　5　6　7　8　9　10

(4) 2　4　6　8　10

(5) 3　2　1　0

Do you know your numbers up to 10? Good job!

© Kumon Publishing Co., Ltd.

4 Numbers up to 10

Date   /   /

Name

/100

1 Which side has more, ⓐ or ⓑ ? Write a ✓ under the box with more.

5 points per question

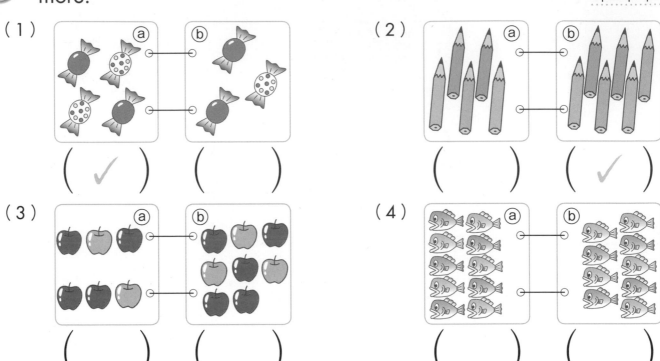

(1)   (✓)   ( )

(2)   ( )   (✓)

(3)   ( )   ( )

(4)   ( )   ( )

2 Which side has more, ⓐ or ⓑ ? Write a ✓ under the box with more.

6 points per question

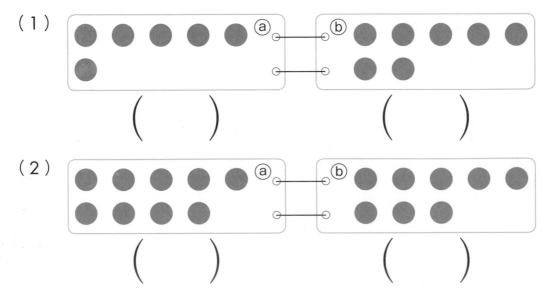

(1)   ( )   ( )

(2)   ( )   ( )

 © Kumon Publishing Co., Ltd.

3 Which is more? Write a ✓ under the larger number. 5 points per question

(1)

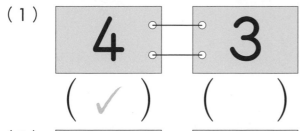

$(\ ✓\ )$ $(\ \ )$

(2)

$(\ \ )$ $(\ \ )$

(3)

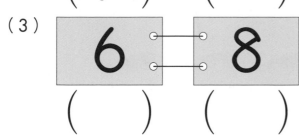

$(\ \ )$ $(\ \ )$

(4)

$(\ \ )$ $(\ \ )$

4 Write the number that is 1 more than the number in the box on the left. 6 points per question

(1)

(Left) (Right)

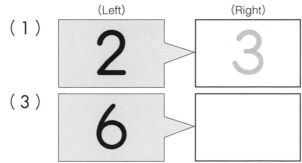

2 ▷ 3

(2)

1 ▷

(3)

6 ▷

(4)

9 ▷

5 Write the number that is 1 less than the number in the box on the left. 6 points per question

(1)

(Left) (Right)

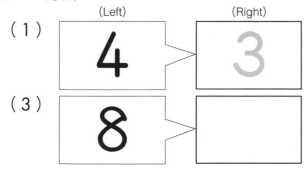

4 ▷ 3

(2)

5 ▷

(3)

8 ▷

(4)

10 ▷

You're doing a great job—keep it up!

© Kumon Publishing Co., Ltd. 9

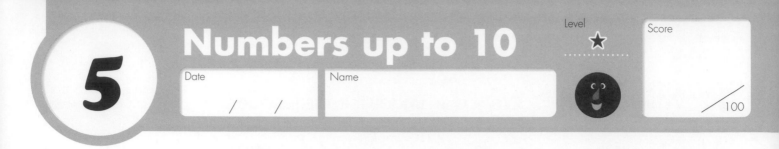

1 7 children wearing hats are standing in line. Answer the questions below.

10 points per question

(1) Color the hats of the first **3** children in line.

Front

(2) Color the hat of the third child from the front.

Front

2 6 cars are waiting in line. Answer the questions below.

10 points per question

(1) Color the first **4** cars in line.

Front

(2) Color the fourth car from the front.

Front

© Kumon Publishing Co., Ltd.

3 There are 7 circles. Answer the questions below.

10 points per question

(1) Color the first **5** circles from the left.

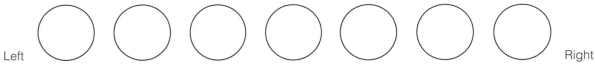

Left   Right

(2) Color the fifth circle from the left.

Left   Right

4 There are 6 cups. Answer the questions below.

10 points per question

| Bob | Cathy | Andy | David | Ellen | Ginny |

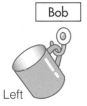     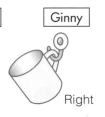

Left   Right

(1) What number cup from the left is Andy's?

(Third)

(2) What number cup from the right is Ellen's?

( )

(3) Whose cup is the third from the right?

( )

(4) Whose cup is the second from the left?

( )

Don't forget!

When you are talking about the order of objects, you use **ordinal numbers**.
Ordinal numbers are place numbers that indicate a distance from a starting point.

· first → second → third → fourth → fifth → sixth → seventh → eighth → ninth → tenth → ……

Numbers can be fun, right?

© Kumon Publishing Co., Ltd.

6 Numbers up to 10

Level ★★

Score

/100

Date   /   /

Name

1 Draw circles until there are 5 in all. The first two questions have hints.

3 points per question

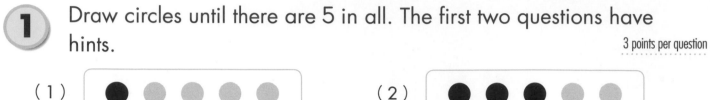

(1)   (2)

(3)   (4)

2 Draw circles until there are 5 in the two boxes combined.

3 points per question

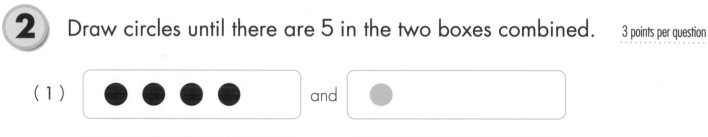

(1)   and

(2)   and

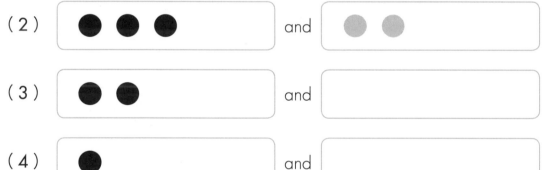

(3)   and

(4)   and

3 Write a number in each box on the right so that the number of circles plus that number equals 5.

3 points per question

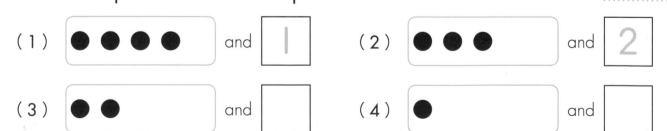

(1)   and   1   (2)   and   2

(3)   and   (4)   and

 © Kumon Publishing Co., Ltd.

④ Write the correct number in each box to complete the sentence.

3 points per question

(1) **5** is **3** and 2

(2) **5** is **1** and ▢

(3) **5** is **4** and ▢

(4) **5** is **2** and ▢

(5) **5** is 3 and **2**

(6) **5** is ▢ and **4**

(7) **5** is ▢ and **1**

(8) **5** is ▢ and **3**

⑤ Write the correct number in each box to complete the sentence.

4 points per question

(1) 5 is 2 and ▢

(2) 5 is ▢ and 3

(3) 5 is 3 and ▢

(4) 5 is ▢ and 2

(5) ▢ is 2 and 3

(6) 5 is 4 and ▢

(7) 5 is ▢ and 1

(8) 5 is 1 and ▢

(9) 5 is ▢ and 4

(10) ▢ is 1 and 4

This is tricky—do your best!

© Kumon Publishing Co., Ltd.

1 Draw circles until there are 6 in all. The first two questions have hints.

2 points per question

(1)

(2)

(3)

(4)

(5)

2 Draw circles until there are 6 in the two boxes combined.

3 points per question

(1) ●●●●● and ●

(2) ●●●● and ●●

(3) ●●● and

(4) ●● and

(5) ● and

© Kumon Publishing Co., Ltd.

3 Write a number in each box on the right so that the number of circles plus that number equals 6.

3 points per question

(1) ● ● ● ● ● and [1]

(2) ● ● ● ● and [ ]

(3) ● ● ● and [ ]

(4) ● ● and [ ]

(5) ● and [ ]

4 Write the correct number in each box to complete the sentence.

4 points per question

(1) 6 is 4 and 2

(2) 6 is 3 and [ ]

(3) 6 is 1 and [ ]

(4) 6 is 5 and [ ]

(5) 6 is 2 and 4

(6) 6 is [ ] and 3

(7) 6 is [ ] and 5

(8) 6 is [ ] and 1

(9) 6 is [ ] and 2

(10) 6 is [ ] and 4

5 Write the correct number in each box to complete the sentence.

4 points per question

(1) 6 is 1 and [ ]

(2) 6 is 2 and [ ]

(3) 6 is 3 and [ ]

(4) 6 is [ ] and 5

(5) [ ] is 4 and 2

Getting the hang of it? Good!

© Kumon Publishing Co., Ltd. 15

1 Draw circles until there are 7 in all. The first two questions have hints.

3 points per question

(1)

(2)

(3)

(4)

(5)

(6)

2 Draw circles until there are 7 in the two boxes combined.

3 points per question

(1)  and

(2) and

(3) and

(4) and

(5) and

(6) and

© Kumon Publishing Co., Ltd.

3 Write a number in each box on the right so that the number of circles plus that number equals 7.

2 points per question

(1) ● ● ● ● ● ● and [1]

(2) ● ● ● ● ● ● and [ ]

(3) ● ● ● ● and [ ]

(4) ● ● ● and [ ]

(5) ● ● and [ ]

(6) ● and [ ]

4 Write the correct number in each box to complete the sentence.

3 points per question

(1) 7 is 6 and [ ]

(2) 7 is 4 and [ ]

(3) 7 is 1 and [ ]

(4) 7 is 3 and [ ]

(5) 7 is 2 and [ ]

(6) 7 is 5 and [ ]

(7) 7 is [ ] and 2

(8) 7 is [ ] and 4

(9) 7 is [ ] and 5

(10) 7 is [ ] and 1

(11) 7 is [ ] and 3

(12) 7 is [ ] and 6

5 Write the correct number in each box to complete the sentence.

4 points per question

(1) 7 is 5 and [ ]

(2) 7 is 3 and [ ]

(3) 7 is [ ] and 3

(4) [ ] is 1 and 6

Now you're getting good!

© Kumon Publishing Co., Ltd. 17

9 Numbers up to 10

Level ★★

Date / /

Name

Score /100

1 Draw circles until there are 8 in all. The first two questions have hints.

2 points per question

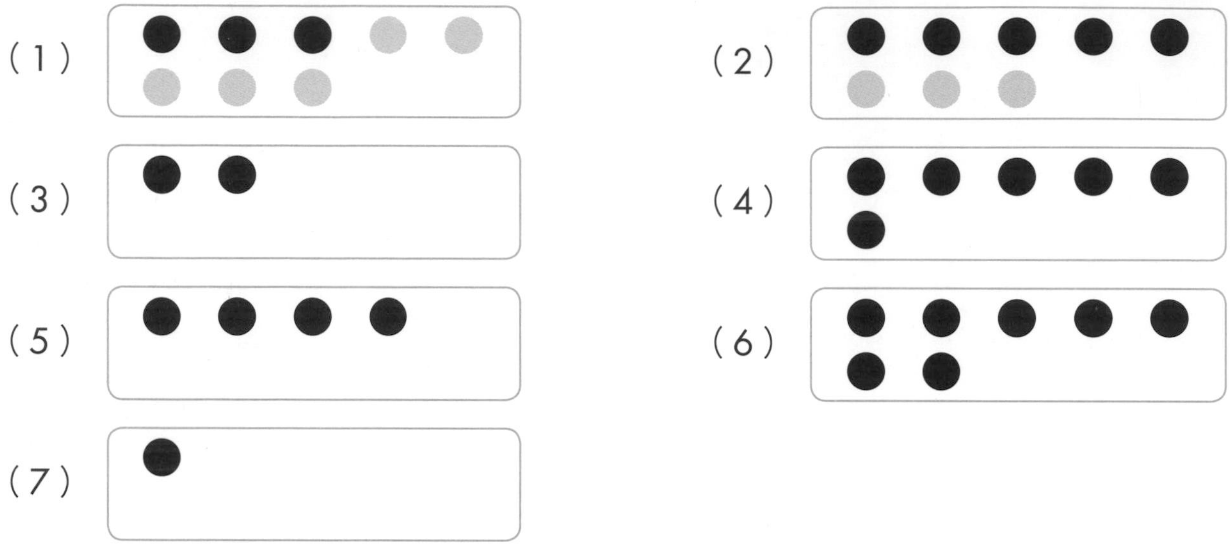

(1)
(2)
(3)
(4)
(5)
(6)
(7)

2 Draw circles until there are 8 in the two boxes combined.

2 points per question

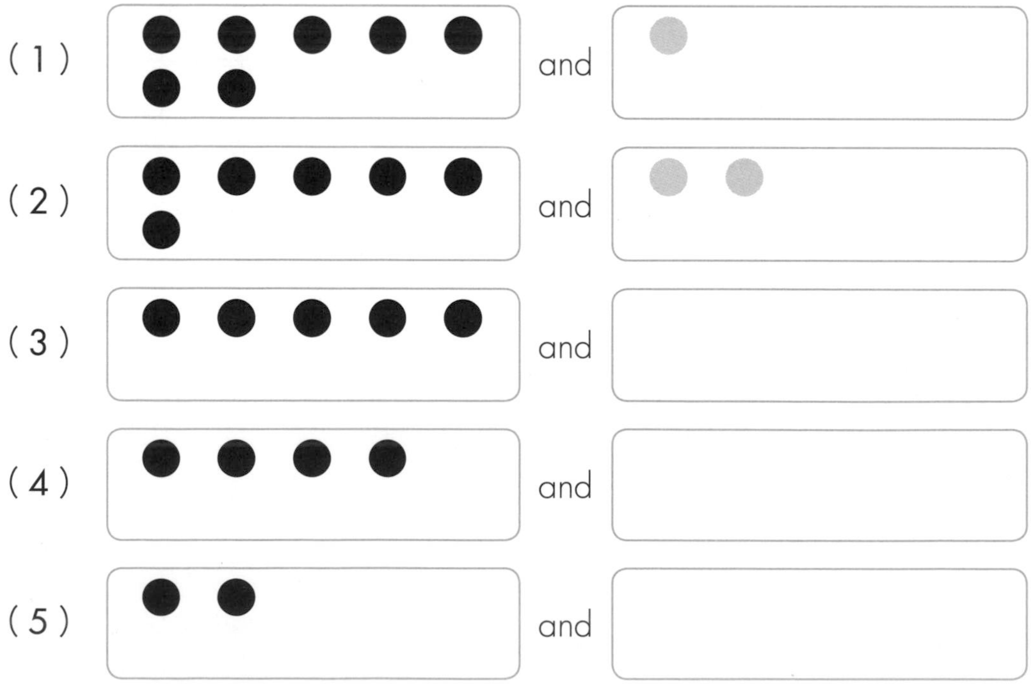

(1) and
(2) and
(3) and
(4) and
(5) and

 © Kumon Publishing Co., Ltd.

3 Write a number in each box on the right so that the number of circles plus that number equals 8.

3 points per question

(1) [●] and [7]

(2) [● ●] and [ ]

(3) [● ● ●] and [ ]

(4) [● ● ● ●] and [ ]

(5) [● ● ● ● ●] and [ ]

(6) [● ● ● ● ● ● ●] and [ ]

4 Write the correct number in each box to complete the sentence.

3 points per question

(1) 8 is 2 and 6

(2) 8 is 3 and [ ]

(3) 8 is 4 and [ ]

(4) 8 is 6 and [ ]

(5) 8 is 7 and [ ]

(6) 8 is 1 and [ ]

(7) 8 is 5 and [ ]

(8) 8 is [ ] and 2

(9) 8 is [ ] and 3

(10) 8 is [ ] and 7

(11) 8 is [ ] and 1

(12) 8 is [ ] and 5

(13) 8 is [ ] and 2

(14) 8 is [ ] and 6

5 Write the correct number in each box to complete the sentence.

4 points per question

(1) 8 is 6 and [ ]

(2) 8 is [ ] and 3

(3) [ ] is 2 and 6

(4) [ ] is 4 and 4

No problem, right?

© Kumon Publishing Co., Ltd. 19

1 Draw circles until there are 9 in all. The first two questions have hints.

2 points per question

(1)

(2)

(3)

(4)

(5)

(6)

(7)

(8)

2 Draw circles until there are 9 in the two boxes combined.

2 points per question

(1)  and

(2) and

(3) and

(4) and

(5) and

(6) and

(7) and

© Kumon Publishing Co., Ltd.

3 Write a number in each box on the right so that the number of circles plus that number equals 9.

3 points per question

(1) ● and 8

(2) ● ● ● and ☐

(3) ● ● ● ● and ☐

(4) ● ● ● ● ● ● and ☐

(5) ● ● ● ● ● ● ● ● and ☐

(6) ● ● ● ● ● ● ● ● and ☐

4 Write the correct number in each box to complete the sentence.

3 points per question

(1) 9 is 2 and ☐

(2) 9 is 5 and ☐

(3) 9 is 8 and ☐

(4) 9 is 4 and ☐

(5) 9 is 1 and ☐

(6) 9 is 3 and ☐

(7) 9 is 6 and ☐

(8) 9 is 7 and ☐

(9) 9 is ☐ and 3

(10) 9 is ☐ and 4

(11) 9 is ☐ and 8

(12) 9 is ☐ and 6

5 Write the correct number in each box to complete the sentence.

4 points per question

(1) 9 is 2 and ☐

(2) 9 is ☐ and 5

(3) 9 is ☐ and 2

(4) ☐ is 8 and 1

Are you almost ready for something else?

11 Numbers up to 10

Level ★★

Date / /   Name

Score /100

1 Draw circles until there are 10 in all. The first two questions have hints.

2 points per question

(1) 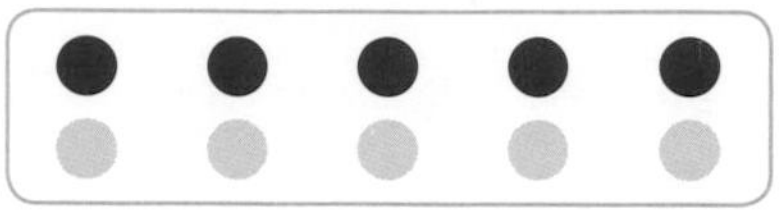

(2) 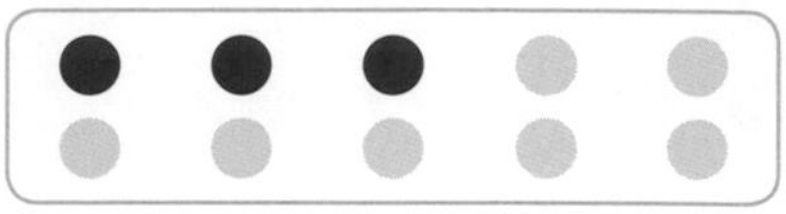

(3)

(4) 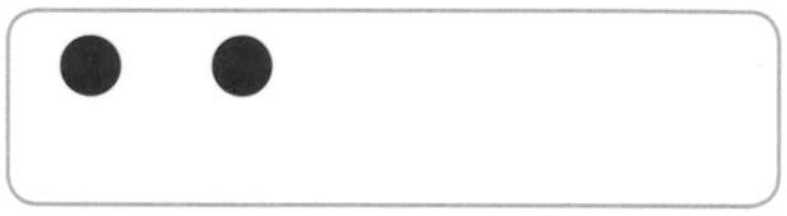

(5) 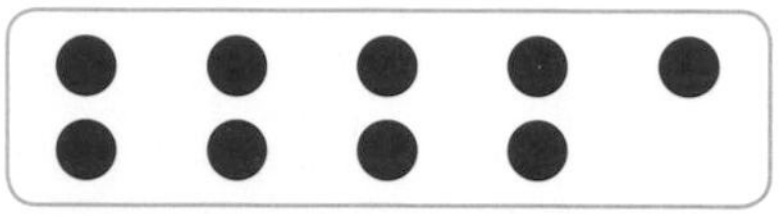

(6) 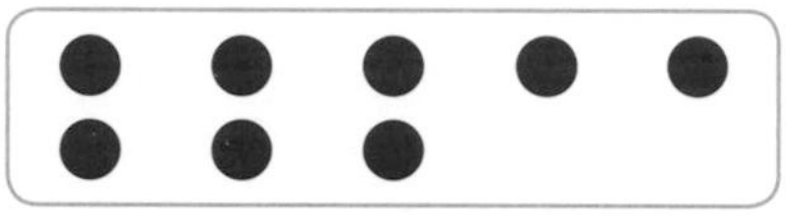

(7) 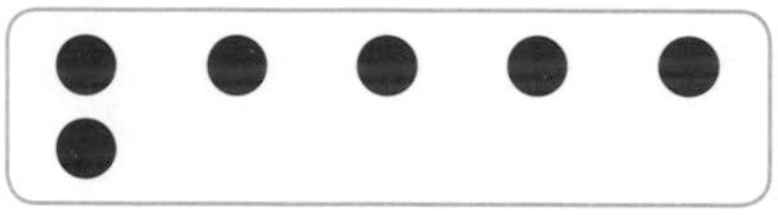

(8) 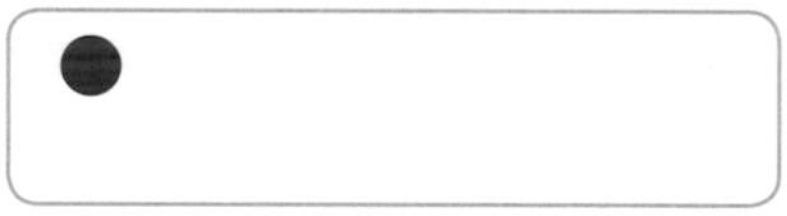

(9)

2 Draw circles until there are 10 in the two boxes combined.

4 points per question

(1) 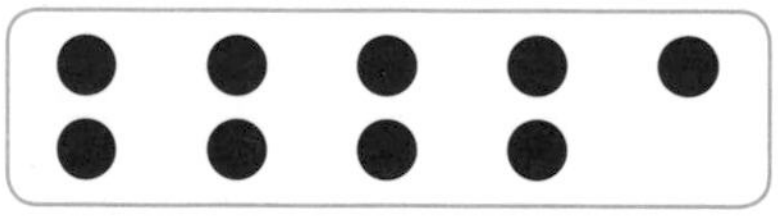 and

(2) 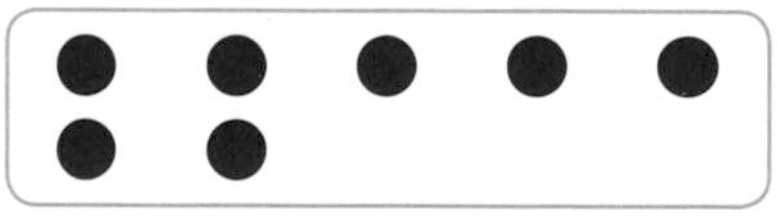 and

(3) 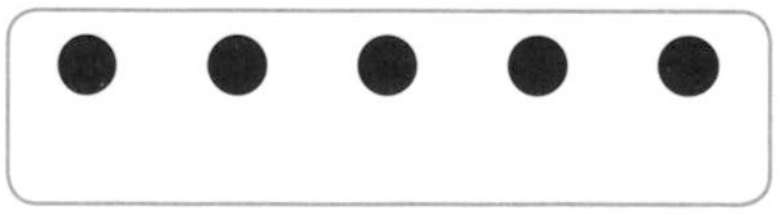 and

(4) and

(5) 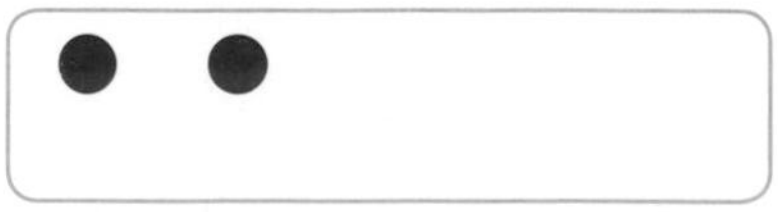 and

© Kumon Publishing Co., Ltd.

3 Write a number in each box on the right so that the number of circles plus that number equals 10.

2 points per question

(1) [●] and [9]

(2) [● ●] and [   ]

(3) [● ● ●] and [   ]

(4) [● ● ● ●] and [   ]

(5) [● ● ● ● ●] and [   ]

(6) [● ● ● ● ● / ●] and [   ]

(7) [● ● ● ● ● / ● ●] and [   ]

(8) [● ● ● ● ● / ● ● ● ●] and [   ]

4 Write the correct number in each box to complete the sentence.

3 points per question

(1) 10 is 3 and [   ]

(2) 10 is 5 and [   ]

(3) 10 is 8 and [   ]

(4) 10 is 4 and [   ]

(5) 10 is 7 and [   ]

(6) 10 is [   ] and 1

(7) 10 is [   ] and 2

(8) 10 is 6 and [   ]

(9) 10 is 9 and [   ]

(10) 10 is [   ] and 3

5 Write the correct number in each box to complete the sentence.

4 points per question

(1) 10 is 2 and [   ]

(2) 10 is 7 and [   ]

(3) [   ] is 1 and 9

(4) [   ] is 4 and 6

Now let's try some new numbers!

© Kumon Publishing Co., Ltd. 23

Numbers up to 20

Level ★ ★

Score /100

1   How many sticks are there? Write the number in each box.

3 points per box

  |   | 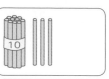 |  |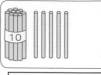

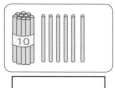 | 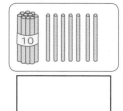 | 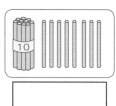 | 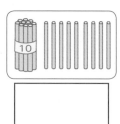 |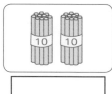

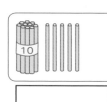 | 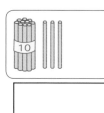 | 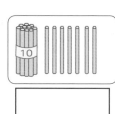 | 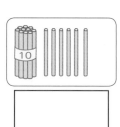 |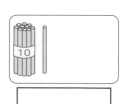

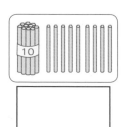 | 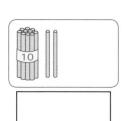 | 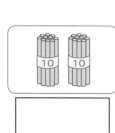 | 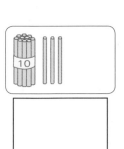 |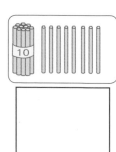

 © Kumon Publishing Co., Ltd.

2 How many objects are in each picture? Write the number in each box.

5 points per question

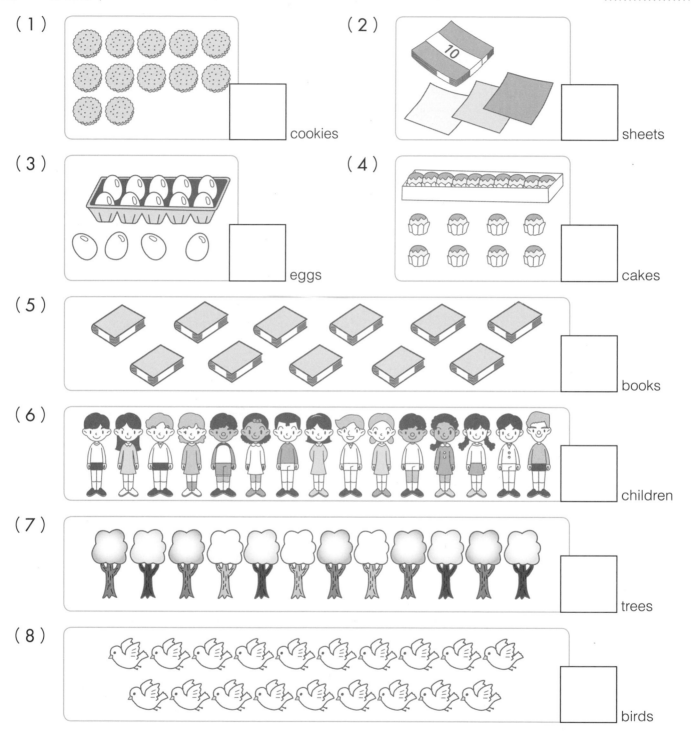

(1)

cookies

(2)

sheets

(3)

eggs

(4)

cakes

(5)

books

(6)

children

(7)

trees

(8)

birds

Do you like to count things? It is fun, right?

13 Numbers up to 20

Date / /　Name

Level ★★★　Score /100

1 Write the correct number in each box to complete the sentence.

2 points per question

(1) 11 is 10 and ☐

(2) 12 is 10 and ☐

(3) 13 is 10 and ☐

(4) 14 is 10 and ☐

(5) 15 is 10 and ☐

(6) 16 is 10 and ☐

(7) 17 is 10 and ☐

(8) 18 is 10 and ☐

(9) 19 is 10 and ☐

(10) 20 is 10 and ☐

2 Write the correct number in each box to complete the sentence.

2 points per question

(1) 11 is ☐ and 1

(2) 12 is ☐ and 2

(3) 13 is ☐ and 3

(4) 14 is ☐ and 4

(5) 15 is ☐ and 5

(6) 16 is ☐ and 6

(7) 17 is ☐ and 7

(8) 18 is ☐ and 8

(9) 19 is ☐ and 9

(10) 20 is ☐ and 10

 © Kumon Publishing Co., Ltd.

3 Write the correct number in each box to complete the sentence.

3 points per question

(1) 10 and 1 is ☐   (2) 10 and 2 is ☐

(3) 10 and 3 is ☐   (4) 10 and 4 is ☐

(5) 10 and 5 is ☐   (6) 10 and 6 is ☐

(7) 10 and 7 is ☐   (8) 10 and 8 is ☐

(9) 10 and 9 is ☐   (10) 10 and 10 is ☐

4 Write the correct number in each box to complete the sentence.

3 points per question

(1) 15 is 10 and ☐   (2) 17 is 10 and ☐

(3) 18 is 10 and ☐   (4) 14 is 10 and ☐

(5) 20 is 10 and ☐

5 Write the correct number in each box to complete the sentence.

3 points per question

(1) 10 and 6 is ☐   (2) 10 and 2 is ☐

(3) 10 and 9 is ☐   (4) 10 and 1 is ☐

(5) 10 and 3 is ☐

You are getting good at addition!

© Kumon Publishing Co., Ltd.   27

Numbers up to 20

Date / /   Name

Level ★★

Score /100

---

1 Write the missing number in each box.

8 points per question

(1)

1	2	3	4	5	6	7	8	9	10
11	12		14	15	16	17	18		20

(2)

1	2	3	4	5	6	7	8	9	10
		13			16		18		

(3)

1	2		4		6	7		9	
	12	13		15					20

(4)

1	2		4	5		7	8		
		13			16			19	20

(5)

1	2			5			8		
11								19	

© Kumon Publishing Co., Ltd.

2 Write the missing number in each box.

6 points per question

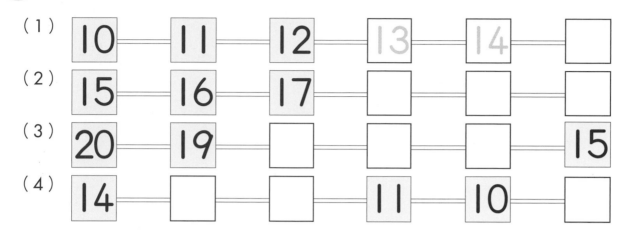

(1) 10   11   12   13   14   ☐

(2) 15   16   17   ☐   ☐   ☐

(3) 20   19   ☐   ☐   ☐   15

(4) 14   ☐   ☐   11   10   ☐

3 Write the missing number in each box.

4 points per question

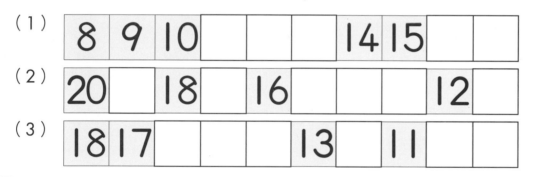

(1) 8   9   10   ☐   ☐   ☐   14   15   ☐   ☐

(2) 20   ☐   18   ☐   16   ☐   ☐   12   ☐

(3) 18   17   ☐   ☐   ☐   13   ☐   11   ☐

4 Write the missing number in each box.

6 points per question

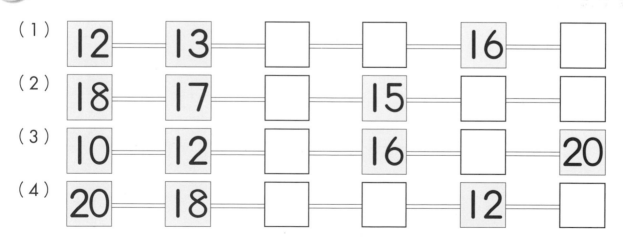

(1) 12   13   ☐   ☐   16   ☐

(2) 18   17   ☐   15   ☐   ☐

(3) 10   12   ☐   16   ☐   20

(4) 20   18   ☐   ☐   12   ☐

Make sure to check the pattern in each number line before filling in the missing numbers!

© Kumon Publishing Co., Ltd.

1 Fill in the missing number in each box on the number lines below.

1 point per box

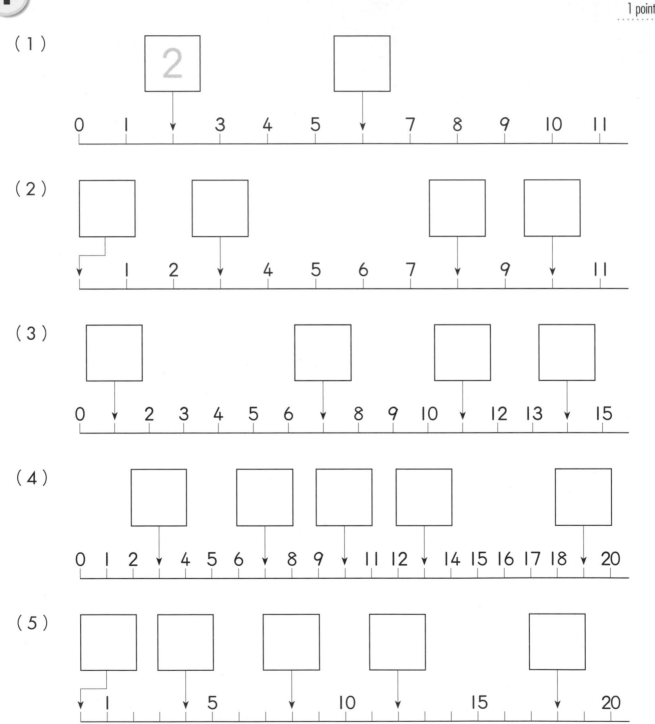

© Kumon Publishing Co., Ltd.

2 Fill in the missing number in each box on the number lines below.

2 points per box

(1)

Boxes: 5, 7, 12, 15

4 | 6 | 8 | 9 | 10 | 11 | 13 | 14

(2)

Boxes: 9, 13, 16, 18

7 | 8 | 10 | 11 | 12 | 14 | 15 | 17

(3)

Boxes: 2, 4, 6, 8, 10

0 | 1 | 3 | 5 | 7 | 9 | 11 | 12

(4)

Boxes: 2, 4, 6, 8, 10

0 | 1 | 5 | 12

(5)

0 | | 4 | | 8 | | 14 | | 20

1 | 3 | 5 | 7 | 9 | 11 | 13 | 15 | 17 | 19 | 21

(6)

| 2 | 4 | | 10 | | | 18 |

1 | 3 | 5 | 7 | 9 | 11 | 13 | 15 | 17 | 19 | 21

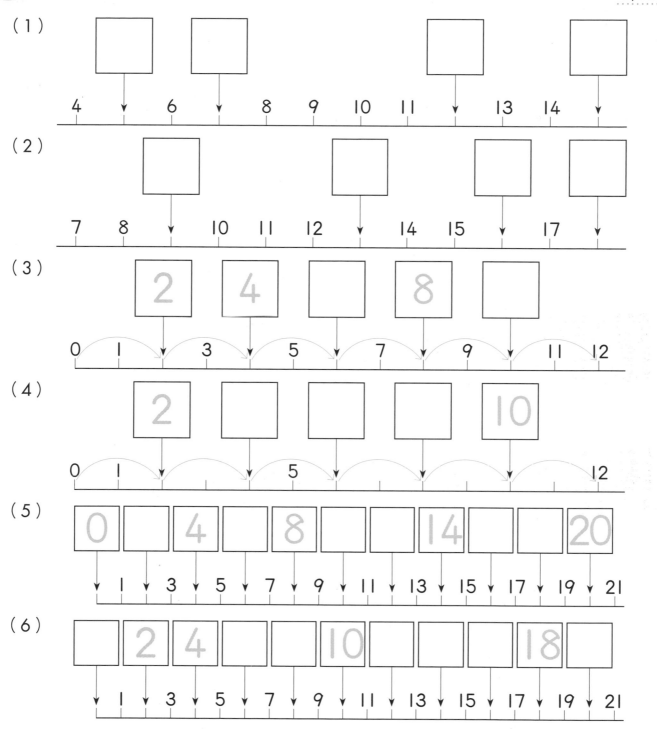

How about a little more practice with numbers up to 20, okay?

© Kumon Publishing Co., Ltd. 31

1 Fill in the box on the right with a number that is 1 more than the number on the left.

5 points per question

6 7 8 9 10 11 12 13 14 15 16 17 18 19 20

(1)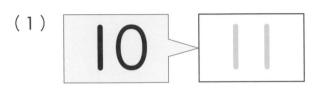

10 ▸ 11

(2) 18 ▸ 19

(3) 12 ▸ ☐

(4) 15 ▸ ☐

(5)

16 ▸ ☐

(6) 19 ▸ ☐

2 Fill in the box on the right with a number that is 1 less than the number on the left.

5 points per question

(1)

11 ▸ 10

(2) 17 ▸ 16

(3)

15 ▸ ☐

(4) 14 ▸ ☐

(5)

12 ▸ ☐

(6) 20 ▸ ☐

© Kumon Publishing Co., Ltd.

3 Using the number line as a guide, fill in each box to complete the sentences below.

5 points per question

0   2 less 5 2 more   10   15   20

(1)   The number that is **2** more than 10 is [ ].

(2)   The number that is **2** less than 10 is [ ].

(3)   The number that is **2** more than 15 is [ ].

(4)   The number that is **2** less than 15 is [ ].

4 Using the number line as a guide, fill in each box to complete the sentences below.

5 points per question

0   3 less 5 3 more   10   15   20

(1)   The number that is **3** more than 10 is [ ].

(2)   The number that is **3** less than 10 is [ ].

(3)   The number that is **3** more than 15 is [ ].

(4)   The number that is **3** less than 15 is [ ].

Remember to use the number line as a hint!

Numbers up to 120

Level ★★

1 How many sticks are there? Write the number in each box.

5 points per question

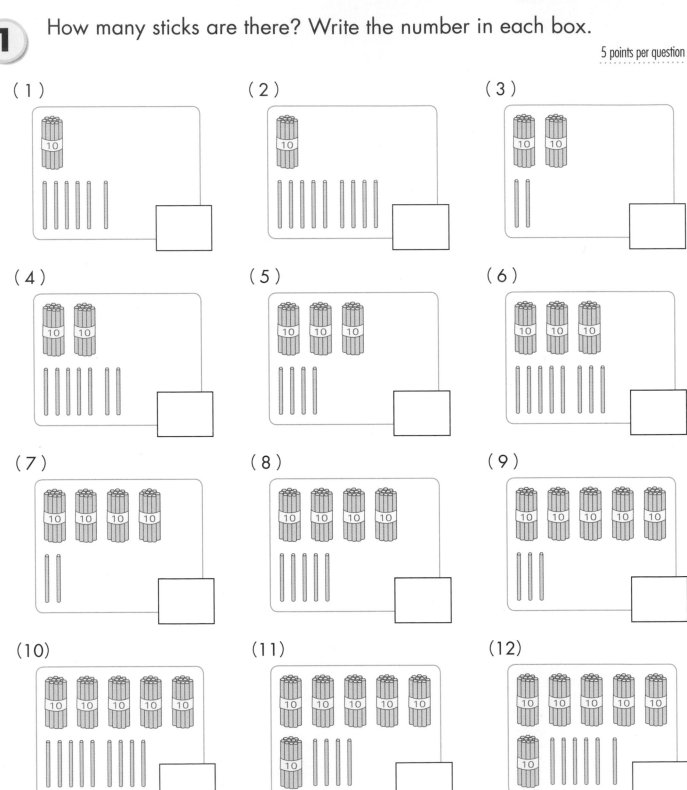

(1)

(2)

(3)

(4)

(5)

(6)

(7)

(8)

(9)

(10)

(11)

(12)

2 How many objects are in each picture? Write the number in each box.

8 points per question

(1)

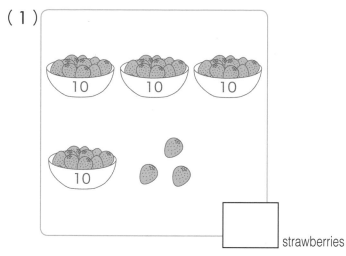

strawberries

(2)

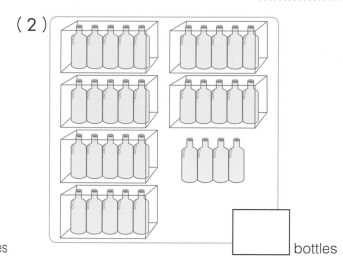

bottles

(3)

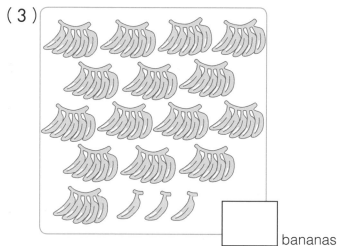

bananas

(4)

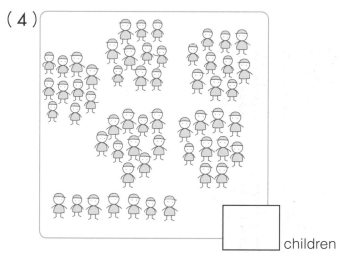

children

(5)

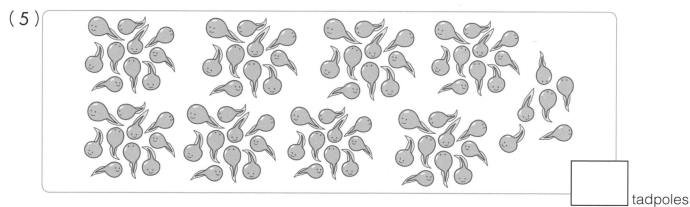

tadpoles

Okay, are you ready for even larger numbers?

18

Date   /   /

Name

Score   /100

1 Write the number in each box to complete the sentence.   5 points per question

(1) 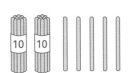   **2** groups of **10** sticks and **5** more sticks are ☐ sticks.

(2) 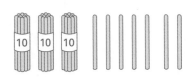   **3** groups of **10** sticks and **7** more sticks are ☐ sticks.

(3)   **5** tens and **7** ones are ☐ .

(4)   **4** tens and **9** ones are ☐ .

(5)   **28** is the number you get after adding ☐ tens and ☐ ones.

(6)   **86** is the number you get after adding ☐ tens and ☐ ones.

(7)   The number you get after adding **5** tens is ☐ .

(8)   The number you get after adding **8** tens is ☐ .

© Kumon Publishing Co., Ltd.

2 Answer the questions below.

5 points per question

2	5
Tens place	Ones place

(1) What number is in the tens place? (   )

(2) What number is in the ones place? (   )

3 Pay attention to the order of the numbers below. Then answer the questions.

(1)(2) 20 points, (3) 10 points

1	2	3	4	5	6	7	8	9	10
11	12	13	14	15	16	17	18	19	20
21	22	23	24	25	26	27	28	29	30
31	32	33	34	35	36	37	38	39	40
41	42	43	44	45	46	47	48	49	50
51	52	53	54	55	56	57	58	59	60
61	62	63	64	65	66	67	68	69	70
71	72	73	74	75	76	77	78	79	80
81	82	83	84	85	86	87	88	89	90
91	92	93	94	95	96	97	98	99	100
101	102	103	104	105	106	107	108	109	110
111	112	113	114	115	116	117	118	119	120

(1) Write ○ on all the numbers that have **7** in the ones place.

(2) Write △ on all the numbers that have **7** in the tens place.

(3) Write all the numbers that have **0** in the ones place below.

(10   )

Wow, this is hard! Good job!

Numbers up to 120

Date   /   /

Name

Score   /100

1 Fill in the missing number in each box on the number lines below.

2 points per box

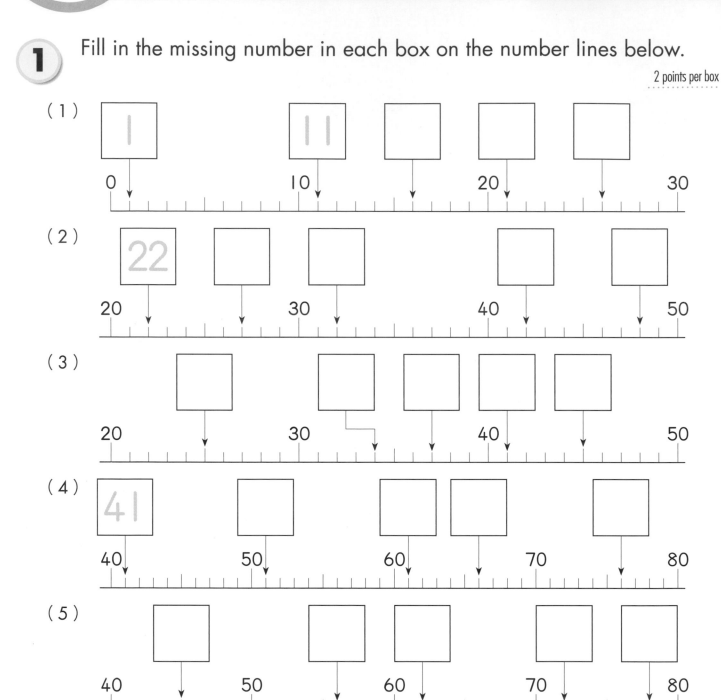

(1)

| 1 | | 11 | | | | |
0   10   20   30

(2)

| 22 | | | | |
20   30   40   50

(3)

20   30   40   50

(4)

| 41 |
40   50   60   70   80

(5)

40   50   60   70   80

(6)

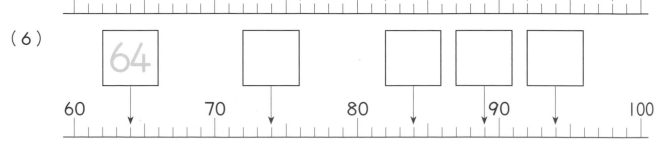

| 64 |
60   70   80   90   100

 © Kumon Publishing Co., Ltd.

2 Using the number line as a guide, fill in each box to complete the sentences below.

4 points per question

```
0              2 less 10  2 more          20                30
|‒|‒|‒|‒|‒|‒|‒|‒|‒|‒|‒|‒|‒|‒|‒|‒|‒|‒|‒|‒|‒|‒|‒|‒|‒|‒|‒|‒|‒|‒|
```

(1) The number that is **2** more than **10** is ☐ .

(2) The number that is **2** less than **10** is ☐ .

(3) The number that is **2** more than **20** is ☐ .

(4) The number that is **2** less than **20** is ☐ .

3 Write the number that is 1 more than the number in the box on the left.

3 points per question

(1) **27** ▷ 28

(2) **38** ▷ ☐

(3) **69** ▷ ☐

(4) **90** ▷ ☐

4 Write the number that is 1 less than the number in the box on the left.

3 points per question

(1) **33** ▷ ☐

(2) **59** ▷ ☐

(3) **81** ▷ ☐

(4) **100** ▷ ☐

You are getting good at these numbers!

1 Which side has more ⓐ or ⓑ ? Write a ✓ in the box next to the picture that has more.

10 points

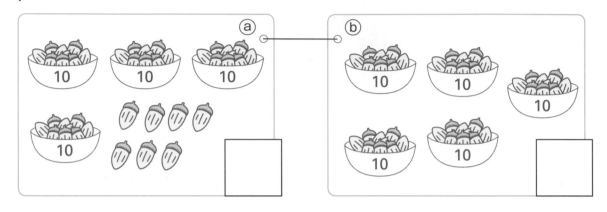

2 Write a ✓ under the side that has more.

5 points per question

(1)

21	19
(✓)	( )

(2)

24	27
( )	( )

(3)

30	33
( )	( )

(4)

46	40
( )	( )

(5)

48	38
( )	( )

(6)

56	66
( )	( )

© Kumon Publishing Co., Ltd.

3 Write a ✓ under the side that has more.   6 points per question

(1) 51 15
(✓) ( )

(2) 28 82
( ) ( )

(3) 79 83
( ) ( )

(4) 64 47
( ) ( )

(5) 99 100
( ) ( )

(6) 89 98
( ) ( )

4 Rearrange each group of numbers from most to least.   6 points per question

(1) 21, 23, 12 ➡ (23, 21, 12)

(2) 9, 19, 90 ➡ ( )

(3) 71, 74, 47 ➡ ( )

(4) 96, 69, 89 ➡ ( )

Now let's try something different.

© Kumon Publishing Co., Ltd.   41

Telling Time

1   What time is it? Write the time under each clock.   5 points per question

(1)

(1:00)

(2)

(2:00)

(3)

(　　　)

(4)

(　　　)

(5)

(　　　)

(6)

(　　　)

2   What time is it? Write the time under each clock.   5 points per question

(1)

(　　　)

(2)

(　　　)

(3)

(　　　)

(4)

(　　　)

© Kumon Publishing Co., Ltd.

3 What time is it? Write the time under each clock.

5 points per question

(1)

(9:30)

(2)

(11:30)

(3)

(   )

(4)

(   )

(5)

(   )

(6)

(   )

4 What time is it? Write the time under each clock.

5 points per question

(1)

(   )

(2)

(   )

(3)

(   )

(4)

(   )

It is good to be able to read the time—well done!
If you need more practice, you could try *My Book of Easy Telling Time*, too.

22

Length

Date   /   /

Name

Level
★ ★

Score

/100

1 The pictures below show different ways to compare the length of two pencils. Write a ✓ under the correct ways, and an × under the wrong ways.

25 points for completion

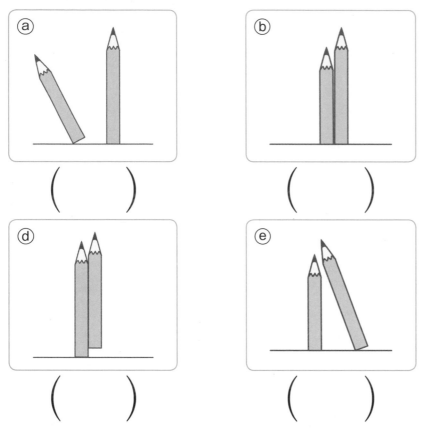

ⓐ ( )

ⓑ ( )

ⓒ ( )

ⓓ ( )

ⓔ ( )

2 Rank the bars below by their length—put a 1 next to the longest bar, and a 2 next to the next-longest bar, for example.

15 points for completion

ⓐ ( )

ⓑ ( )

ⓒ ( )

3 Compare the length of the two lines shown in ⓐ and ⓑ, and write a ✓ next to the longer one.

10 points per question

(1) ⓐ (   )

ⓑ (   )

(2) ⓐ (   )

ⓑ (   )

(3) ⓐ (   )

ⓑ (   )

(4) (   )

ⓐ

ⓑ (   )

(5) (   )

ⓐ

ⓑ (   )

4 Rank the bars below by their length—put a 1 next to the longest bar, and a 2 next to the next-longest bar, for example.

10 points for completion

ⓐ (   )

ⓑ (   )

ⓒ (   )

Can you tell which lines are longer? Well done.

1 Compare the length of the two objects shown in ⓐ and ⓑ, and write a ✓ next to the longer one.

10 points per question

(1)

ⓐ

( )

ⓑ

( )

(2)

ⓐ

( )

ⓑ

( )

2 Compare the length of the two objects shown in ⓐ and ⓑ, and write a ✓ next to the longer one.

10 points per question

(1)

ⓐ

( )

ⓑ

( )

(2)

ⓐ

( )

ⓑ

( )

(3)

ⓐ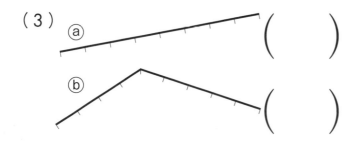

( )

ⓑ

( )

(4)

ⓐ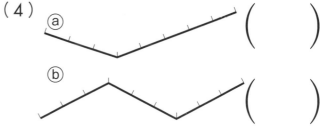

( )

ⓑ

( )

© Kumon Publishing Co., Ltd.

3 Compare the lengths of the items below and then answer the questions.

10 points per question

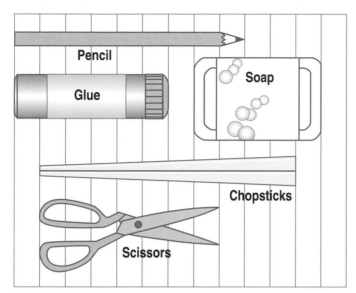

(1) The longest:

(*Pencl*)

(2) The shortest:

( )

4 Compare the lengths of two pencils below and then answer the questions. Each box represents a unit.

10 points per question

(1)

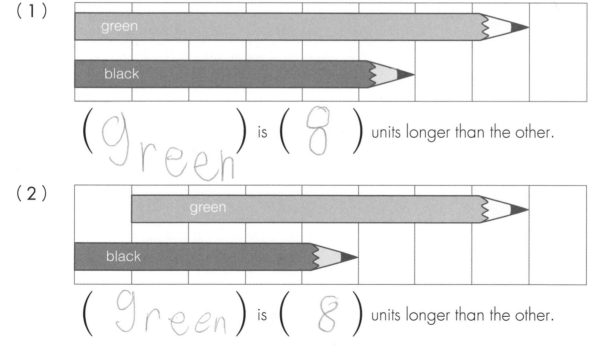

(*green*) is (*8*) units longer than the other.

(2)

(*green*) is (*8*) units longer than the other.

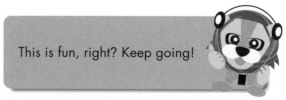

This is fun, right? Keep going!

24 Weight

Level ★★

Date / /

Name

Score /100

1 Which object is heavier? Circle the heavier object.

5 points per question

(1)

(2)

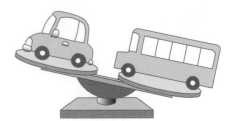

(3)

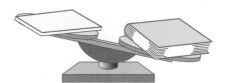

(4)

2 All of the blocks below are the same size and weight. Circle the heavier group of blocks.

5 points per question

(1)   A   B

(2)   A   B

(3)   A   B

(4)   A   B

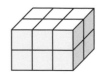

 © Kumon Publishing Co., Ltd.

3 How many cubes are needed on the left side of each scale below in order to balance it?

12 points per question

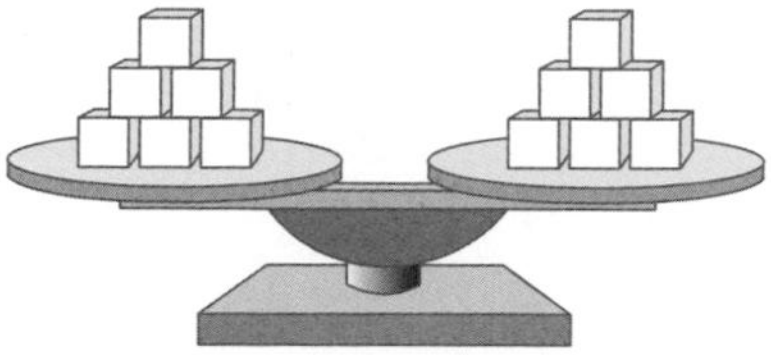

➡ same weight

(1)

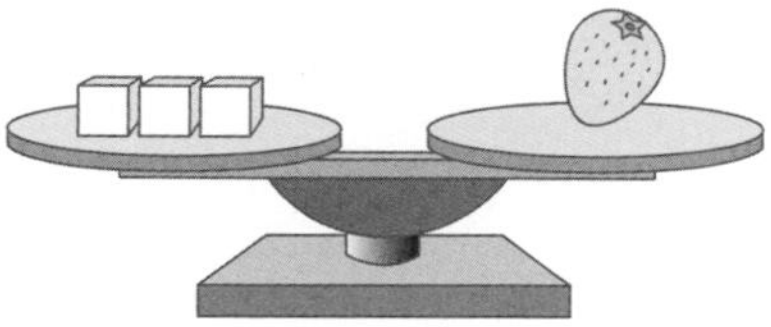

➡ _____3_____ cubes

(2)

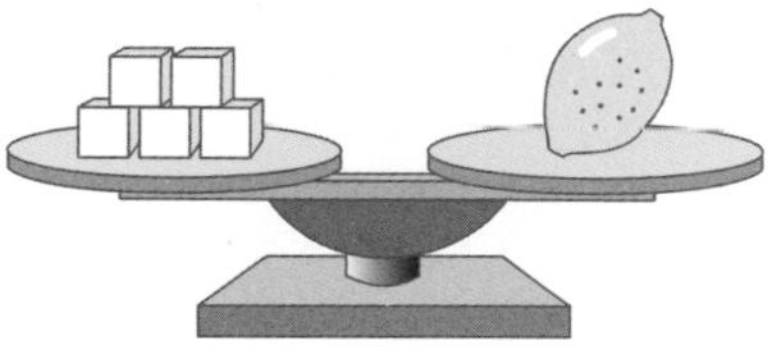

➡ _____ cubes

(3)

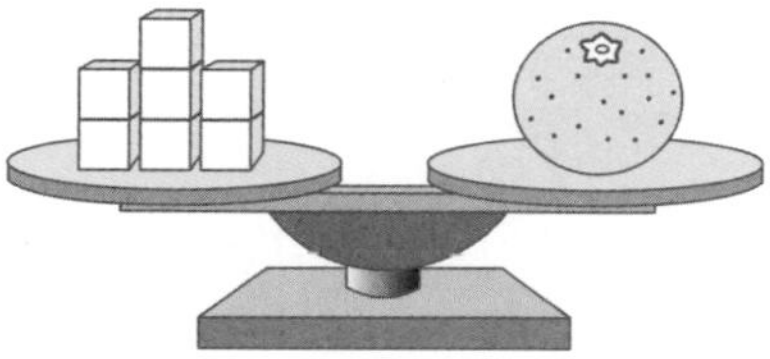

➡ _____ cubes

(4)

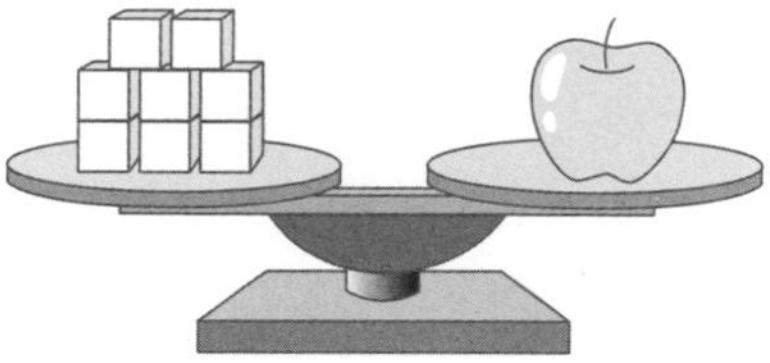

➡ _____ cubes

(5)

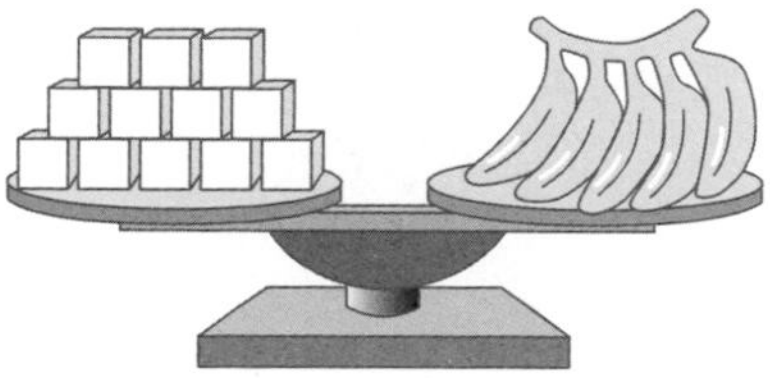

➡ _____ cubes

When the scale is balanced, both sides are the same weight.

25 Area

Date / /

Name

Level ★★

Score /100

1 Which is larger? Circle the larger shape.

6 points per question

(1)

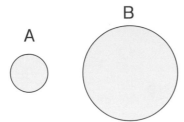

(2)

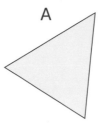

(3)

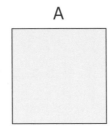

(4)

2 Below are some objects covered by squares of colored paper. Count the squares to answer the questions below.

6 points per question

A   B   C   D   E

(1) Which shape has the largest area?

(   )

(2) Which shape has the smallest area?

(   )

 © Kumon Publishing Co., Ltd.

3 How many squares are used to make each figure below? 8 points per question

(1)

( )

(2)

( )

(3)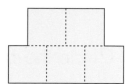

( )

(4)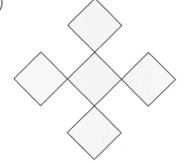

( )

(5)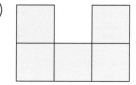

( )

(6)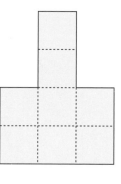

( )

(7)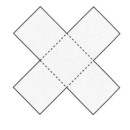

( )

(8)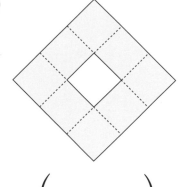

( )

The definition of area is `surface within a set of lines.´ That makes sense, right?

© Kumon Publishing Co., Ltd.

26

Volume

Date

/ /

Name

Level
★ ★

Score

/100

1 Circle the bottle with more milk in it.

8 points per question

(1)

A 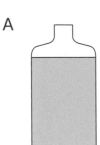   B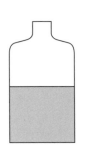

(2)

A 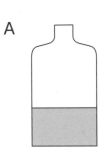   B

(3)

A 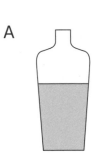   B

(4)

A 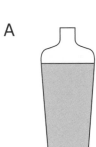   B

2 Circle the bottle with less milk in it.

8 points per question

(1)

A    B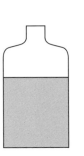

(2)

A 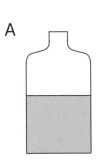   B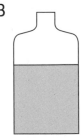

(3)

A 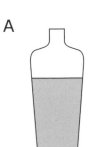   B

(4)

A    B

© Kumon Publishing Co., Ltd.

3 What is the volume of each object on the left?

(1)

_____3_____ cups

(2)

_____ cups

(3)

_____ cups

(4)

_____ cups

(5)

_____ cups

(6)

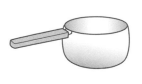

_____ cups

Volume is the space inside an object.
Does that make sense?

1 Look at the thermometers below. Which is warmer, A or B? Circle the warmer one.

8 points per question

(1) A 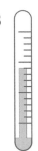 B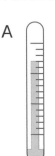

(2) A  B

(3) A  B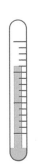

(4) A  B

2 Look at the thermometers below. Which is colder, A or B? Circle the colder one.

8 points per question

(1) A  B

(2) A  B

(3) A  B

(4) A  B

© Kumon Publishing Co., Ltd.

3 Look at the thermometers below. Which is the warmest? Which is the coldest? Circle the warmest one and put a ✓ by the coldest one.

12 points per question

(1) A  B  C 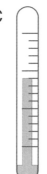 D 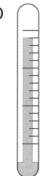 E

(2) A  B 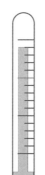 C  D 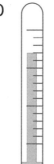 E

(3) A  B  C  D  E

Is it warm or cold today in your house?

© Kumon Publishing Co., Ltd.

Coins

penny		Front I¢	Back I¢

1

Add the value of each row of coins. Then trace the amount in the box on the right.

4 points per question

(1)   🪙 | I | ¢   I penny

(2)   🪙 | I | ¢   I penny

(3)   🪙 🪙 | 2 | ¢   2 pennies

(4)   🪙 🪙 | 2 | ¢   2 pennies

(5)   🪙 🪙 | 2 | ¢   2 pennies

2

Add the value of each row of coins. Then write the amount in the box on the right.

3 points per question

(1)   🪙   [ ] ¢   I penny

(2)   🪙 🪙   [ ] ¢   2 pennies

(3)   🪙 🪙 🪙   [ ] ¢   3 pennies

(4)   🪙 🪙 🪙 🪙   [ ] ¢   4 pennies

(5)   🪙 🪙 🪙 🪙 🪙   [ ] ¢   5 pennies

(6)   🪙 🪙 🪙 🪙 🪙 🪙   [ ] ¢   6 pennies

(7)   🪙 🪙 🪙 🪙 🪙 🪙 🪙   [ ] ¢   7 pennies

(8)   🪙 🪙 🪙 🪙 🪙 🪙 🪙 🪙   [ ] ¢   8 pennies

(9)   🪙 🪙 🪙 🪙 🪙 🪙 🪙 🪙 🪙   [ ] ¢   9 pennies

(10)   🪙 🪙 🪙 🪙 🪙 🪙 🪙 🪙 🪙 🪙   [ ] ¢   I0 pennies

© Kumon Publishing Co., Ltd.

nickel	Front 5¢	Back 5¢

3 Add the value of each row of coins. Then trace the amount in the box on the right.

4 points per question

(1) 🪙   **5** ¢   1 nickel

(2) 🪙   **5** ¢   1 nickel

(3) 🪙🪙   **10** ¢   2 nickels

(4) 🪙🪙   **10** ¢   2 nickels

(5) 🪙🪙   **10** ¢   2 nickels

4 Add the value of each row of coins. Then write the amount in the box on the right.

3 points per question

(1) 🪙   ☐ ¢   1 nickel

(2) 🪙🪙   ☐ ¢   2 nickels

(3) 🪙🪙🪙   ☐ ¢   3 nickels

(4) 🪙🪙🪙🪙   ☐ ¢   4 nickels

(5) 🪙🪙🪙🪙🪙   ☐ ¢   5 nickels

(6) 🪙🪙🪙🪙🪙🪙   ☐ ¢   6 nickels

(7) 🪙🪙🪙🪙🪙🪙🪙   ☐ ¢   7 nickels

(8) 🪙🪙🪙🪙🪙🪙🪙🪙   ☐ ¢   8 nickels

(9) 🪙🪙🪙🪙🪙🪙🪙🪙🪙   ☐ ¢   9 nickels

(10) 🪙🪙🪙🪙🪙🪙🪙🪙🪙🪙   ☐ ¢   10 nickels

Do you have a piggy bank?
Does it have pennies and nickels in it?

29

Coins

Level ★★

Score
/100

Date   /   /

Name

| dime | Front 10¢ | Back 10¢ |

1 Add the value of each row of coins. Then trace the amount in the box on the right.

4 points per question

(1)   🪙   10 ¢   1 dime

(2)   🪙   10 ¢   1 dime

(3)   🪙 🪙   20 ¢   2 dimes

(4)   🪙 🪙   20 ¢   2 dimes

(5)   🪙 🪙   20 ¢   2 dimes

2 Add the value of each row of coins. Then write the amount in the box on the right.

3 points per question

(1)   🪙   ¢   1 dime

(2)   🪙 🪙   ¢   2 dimes

(3)   🪙 🪙 🪙   ¢   3 dimes

(4)   🪙 🪙 🪙 🪙   ¢   4 dimes

(5)   🪙 🪙 🪙 🪙 🪙   ¢   5 dimes

(6)   🪙 🪙 🪙 🪙 🪙 🪙   ¢   6 dimes

(7)   🪙 🪙 🪙 🪙 🪙 🪙 🪙   ¢   7 dimes

(8)   🪙 🪙 🪙 🪙 🪙 🪙 🪙 🪙   ¢   8 dimes

(9)   🪙 🪙 🪙 🪙 🪙 🪙 🪙 🪙 🪙   ¢   9 dimes

(10)   🪙 🪙 🪙 🪙 🪙 🪙 🪙 🪙 🪙 🪙   ¢   10 dimes

© Kumon Publishing Co., Ltd.

quarter	Front 25¢	Back 25¢

3 Add the value of each row of coins. Then trace the amount in the box on the right.

4 points per question

(1) 25 ¢   1 quarter

(2) 25 ¢   1 quarter

(3) 50 ¢   2 quarters

(4) 50 ¢   2 quarters

(5) 50 ¢   2 quarters

4 Add the value of each row of coins. Then write the amount in the box on the right.

3 points per question

(1) ⬜ ¢   1 quarter

(2) ⬜ ¢   2 quarters

(3) ⬜ ¢   3 quarters

(4) ⬜ ¢   4 quarters

(5) ⬜ ¢   5 quarters

(6) ⬜ ¢   6 quarters

(7) ⬜ ¢   7 quarters

(8) ⬜ ¢   8 quarters

(9) ⬜ ¢   9 quarters

(10) ⬜ ¢   10 quarters

Do you know your money now? If not, you could try *My Book of Money: Counting Coins* for more practice!

Coins

1 Add the value of each group of coins. Then write the amount in the box on the right.

5 points per question

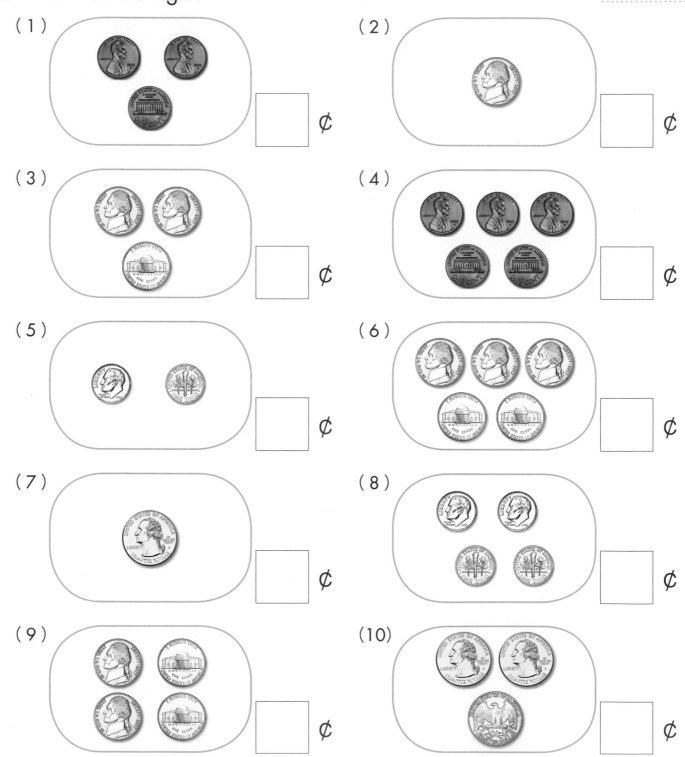

(1) ____ ¢

(2) ____ ¢

(3) ____ ¢

(4) ____ ¢

(5) ____ ¢

(6) ____ ¢

(7) ____ ¢

(8) ____ ¢

(9) ____ ¢

(10) ____ ¢

 © Kumon Publishing Co., Ltd.

2 Add the value of each group of coins. Then write the amount in the box on the right.

5 points per question

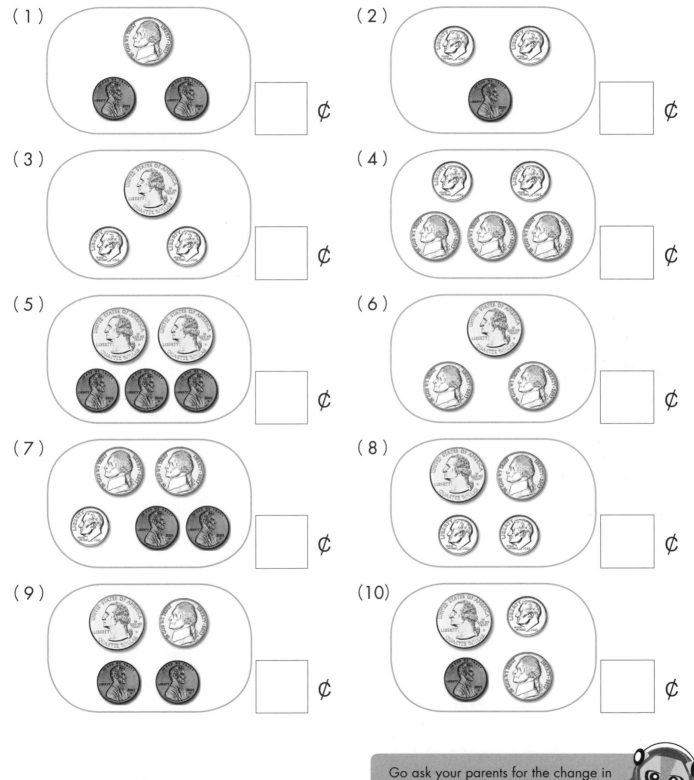

(1) ☐ ¢

(2) ☐ ¢

(3) ☐ ¢

(4) ☐ ¢

(5) ☐ ¢

(6) ☐ ¢

(7) ☐ ¢

(8) ☐ ¢

(9) ☐ ¢

(10) ☐ ¢

Go ask your parents for the change in their pocket—can you tell them how much they have? Good job!

31 Shapes

Date / /

Name

Level ★★★

Score
/100

1 The picture shows some blocks stacked up in a pile. Which block is on top? Find the top block in the blocks on the right, and then write a ✓ under it.

10 points

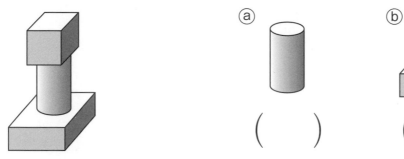

ⓐ

ⓑ

ⓒ

( )   ( )   ( )

2 Write a ✓ under the objects on the right that are similar to the block on the left.

5 points per question

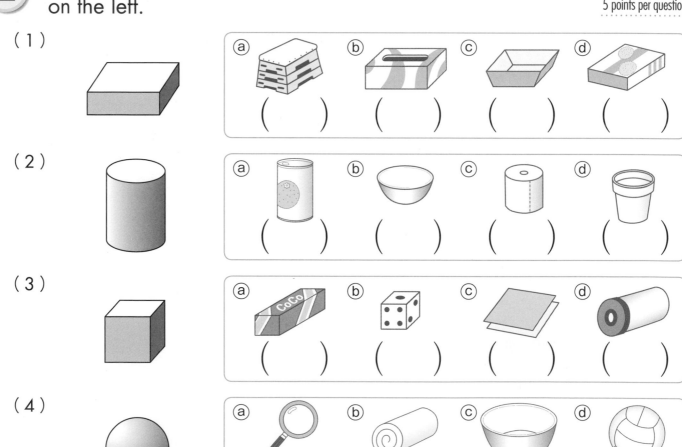

(1)

ⓐ   ⓑ   ⓒ   ⓓ

( )   ( )   ( )   ( )

(2)

ⓐ   ⓑ   ⓒ   ⓓ

( )   ( )   ( )   ( )

(3)

ⓐ   ⓑ   ⓒ   ⓓ

( )   ( )   ( )   ( )

(4)

ⓐ   ⓑ   ⓒ   ⓓ

( )   ( )   ( )   ( )

3 What types of shapes are the objects below? Connect the objects below to shapes ⓐ—ⓒ that are similar.

30 points for completion

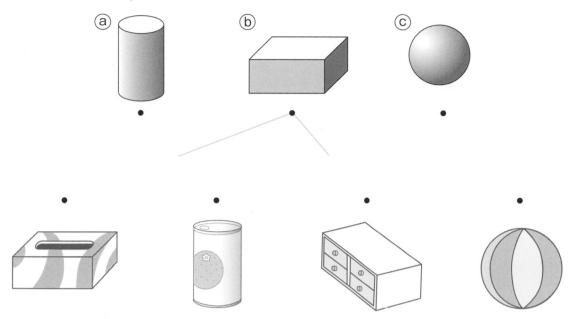

4 The picture below shows a stack of blocks. How many of each block on the right are used in the stack?

10 points per question

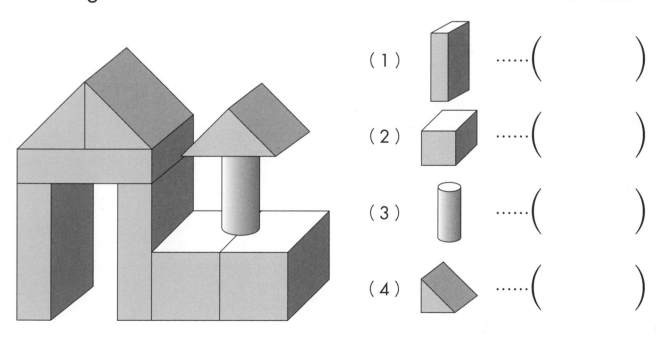

(1) ......(   )

(2) ......(   )

(3) ......(   )

(4) ......(   )

It is fun to build with blocks, right?

1 If you draw a line around the base of each of the blocks on the left, what shape do you get? Write a ✓ under the correct shape.

10 points per question

(1)

ⓐ ☐ ( )   ⓑ ▯ ( )   ⓒ ◯ ( )   ⓓ △ ( )

(2)

ⓐ ☐ ( )   ⓑ ▯ ( )   ⓒ ◯ ( )   ⓓ △ ( )

(3)

ⓐ ☐ ( )   ⓑ ▯ ( )   ⓒ ◯ ( )   ⓓ △ ( )

2 If you draw a line around the base of the blocks in ⓐ—ⓓ, what shapes do you get? Draw a line to connect ⓐ—ⓓ to the correct shapes.

20 points for completion

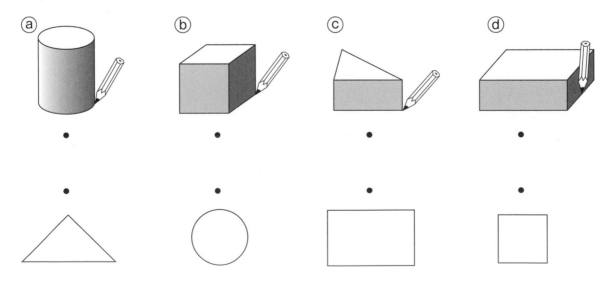

3 What shapes do you see in this figure? Write a ✓ under the correct shapes.

5 points

ⓐ △   ⓑ ▭   ⓒ ▢

( )   ( )   ( )

4 Write a ✓ under the shapes that are on the bases and the faces of these two blocks on the left.

10 points per question

(1)

ⓐ ◯   ⓑ △   ⓒ ▢   ⓓ ▭

( )   ( )   ( )   ( )

(2)

ⓐ ◯   ⓑ △   ⓒ ▢   ⓓ ▭

( )   ( )   ( )   ( )

5 Connect the blocks ⓐ, ⓑ to the shapes on their bases and faces.

25 points for completion

ⓐ   ⓑ

Look at all the shapes around you right now —shapes are everywhere!

© Kumon Publishing Co., Ltd.

Shapes

Date   /   /

Name

Level ★★★

Score   /100

1 How many times is used in the shapes below?

5 points per question

(1)

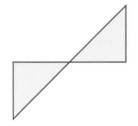

(2)

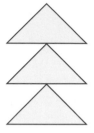

(3)

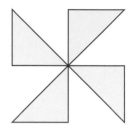

(   )   (   )   (   )

2 These shapes use  two times. Draw a line to divide each shape into two pieces of .

5 points per question

(1)

(2)

(3)

3 These shapes use 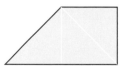 three times. Draw two lines in order to divide each shape into three pieces of .

5 points per question

(1)

(2)

(3)

(4)

© Kumon Publishing Co., Ltd.

4   How many times is used in the shapes below?

10 points per question

(1)

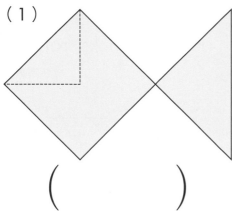

(   )

(2)

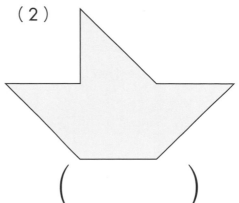

(   )

5   How many times is used in the shapes below?

10 points per question

(1)

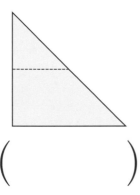

(   )

(2)

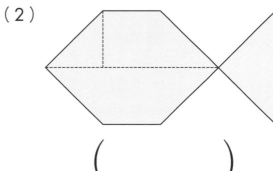

(   )

(3)

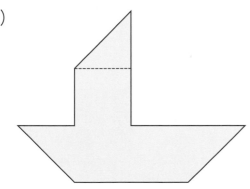

(   )

You can make a lot of shapes with a triangle!

© Kumon Publishing Co., Ltd.

Level ★★★

Date / /

Name

Score /100

1 What type of shapes do you see if you look at these blocks from the front? Write a ✓ under each correct shape.

15 points per question

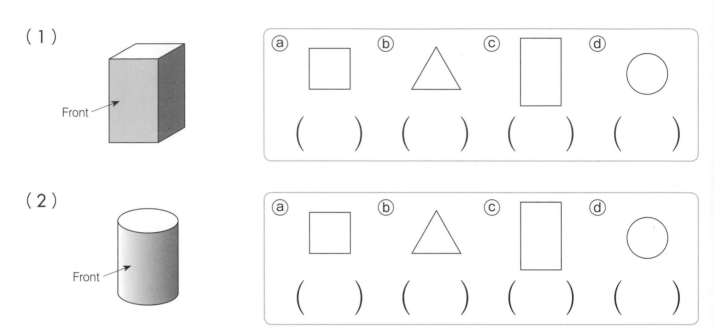

(1)

Front

ⓐ □ ( ) ⓑ △ ( ) ⓒ ▭ ( ) ⓓ ○ ( )

(2)

Front

ⓐ □ ( ) ⓑ △ ( ) ⓒ ▭ ( ) ⓓ ○ ( )

2 What type of shapes do you see if you look at these blocks from the top? Write a ✓ under each correct shape.

15 points per question

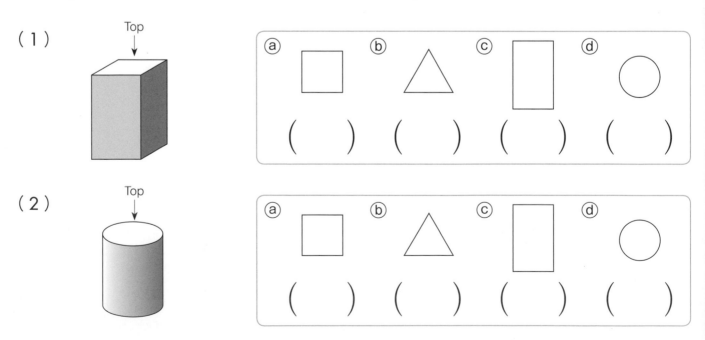

(1)

Top

ⓐ □ ( ) ⓑ △ ( ) ⓒ ▭ ( ) ⓓ ○ ( )

(2)

Top

ⓐ □ ( ) ⓑ △ ( ) ⓒ ▭ ( ) ⓓ ○ ( )

© Kumon Publishing Co., Ltd.

3 What type of shapes do you see if you look at this block from the front and the top? Write a ✓ under the correct combination of shapes on the right.

15 points

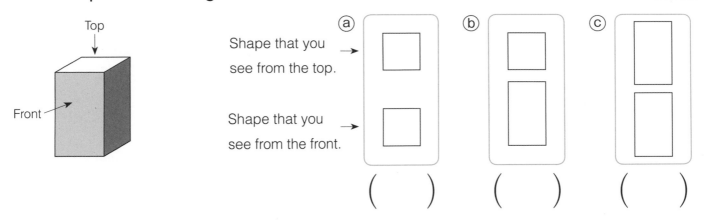

Shape that you see from the top. →

Shape that you see from the front. →

ⓐ   ( )   ⓑ   ( )   ⓒ   ( )

4 Connect blocks ⓐ—ⓓ to the correct combination of shapes below.

25 points for completion

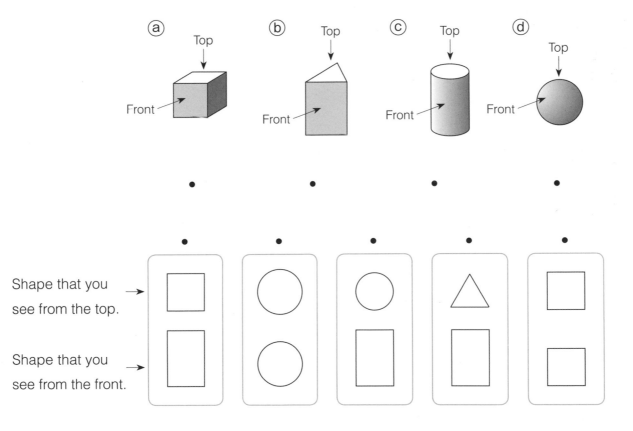

Shape that you see from the top. →

Shape that you see from the front. →

Good job! Now let's review!

Level ★★★

Score   /100

1 How many sticks are there on the left? How many candies on the right? Write the correct amount in the boxes below.

5 points per question

(1)

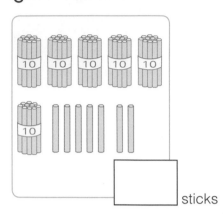

[ ] sticks

(2)

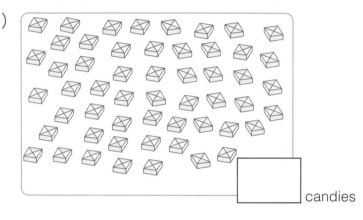

[ ] candies

2 Write the correct number in each box to complete the number sentence.

5 points per box

(1)   **7** tens and **6** ones are [ ] .

(2)   **10** tens are [ ] .

(3)   **68** is the number you get when you add [ ] tens and [ ] ones.

(4)   **80** is the number you get when you add [ ] tens.

(5)   The number in the tens place of **49** is [ ] , and the ones place is [ ] .

(6)   The number with a **0** in the ones place and a **3** in the tens place is [ ] .

3 Which is more? Write a ✓ under the larger number. 10 points per question

(1)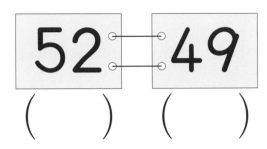

( ) ( )

(2)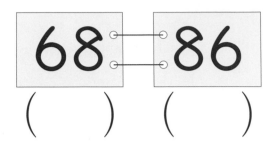

( ) ( )

4 Which is longer? Write a ✓ under the longer object. 10 points per question

(1) ⓐ   ( )

ⓑ   ( )

(2) ⓐ ( )

ⓑ ( )

5 What types of shapes are the blocks below? Connect blocks ⓐ—ⓓ to the object that is similar. 10 points for completion

ⓐ  ⓑ  ⓒ 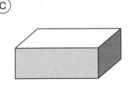 ⓓ

• • • •

• • • •

Almost there! Well done.

36 Review

Level
★★★

Date
/   /

Name

Score
/100

1 Fill in the missing number in each box on the number line below.

2 points per box

(1)

| | 1 | 5 | 10 | 15 | 20 |

(2)

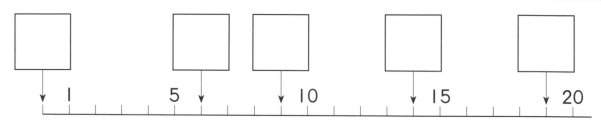

40   50   60   70   80   90   100

2 Rearrange the numbers from most to least.

5 points per question

(1) 34, 33, 43 ➡ (   )

(2) 97, 79, 89 ➡ (   )

3 Write the correct number in each box to complete the number sentence.

5 points per question

(1) The number that is **2** more than **60** is [   ].

(2) The number that is **2** less than **60** is [   ].

(3) The number that is **2** more than **39** is [   ].

(4) The number that is **2** less than **39** is [   ].

4 Write the missing number in each box.

6 points per question

(1) 25 | ☐ | 27 | 28 | ☐ | ☐ | 31 | 32

(2) 78 | 79 | ☐ | 81 | 82 | ☐ | 84 | ☐

(3) 100 | ☐ | 98 | ☐ | 96 | 95 | ☐ | 93

(4) 64 | 66 | ☐ | 70 | ☐ | 74 | 76 | ☐

(5) 23 | 33 | ☐ | 53 | 63 | ☐ | 83 | ☐

5 Write a ✓ under the shapes on the bases and the faces of each block below.

10 points per question

(1)

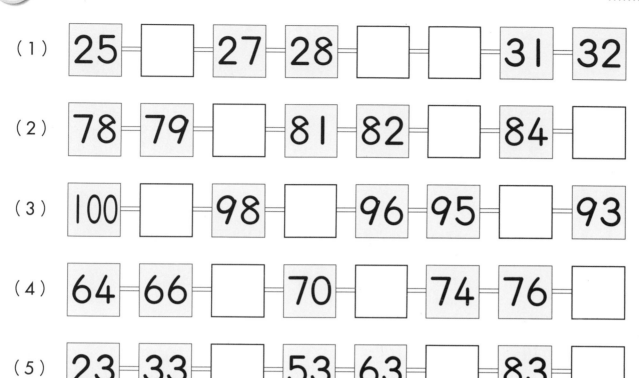

ⓐ ▽ ( )   ⓑ ☐ ( )   ⓒ ▭ ( )   ⓓ ◯ ( )

(2)

ⓐ △ ( )   ⓑ ☐ ( )   ⓒ ▭ ( )   ⓓ ◯ ( )

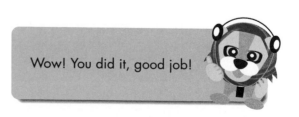

Wow! You did it, good job!

© Kumon Publishing Co., Ltd. 73

1 Numbers up to 10 — pp 2,3

1

(1)
| 1 | 2 | 3 | 4 | 5 |

(2)
| 4 | 3 | 2 | 1 | 0 |

(3)
| 6 | 7 | 8 | 9 | 10 |

(4)
| 10 | 9 | 8 | 7 | 6 |

2
(1) 2 (2) 4 (3) 3 (4) 5 (5) 6
(6) 8 (7) 7 (8) 9 (9) 10 (10) 7

2 Numbers up to 10 — pp 4,5

1 (1) 5 (2) 8 (3) 9 (4) 6

2 (1), (2), (3)

3
| 3 | 2 | 1 | 0 |

4 (1) (From the left) 3, 2, 1, 0
(2) 3, 2, 0   (3) 1, 0, 3

3 Numbers up to 10 — pp 6,7

1
(1) 1 — 2 — 3 — 4 — 5
(2) 6 — 7 — 8 — 9 — 10
(3) 10 — 9 — 8 — 7 — 6
(4) 4 — 3 — 2 — 1 — 0

2
(1) 1 2 3 4 5 6 7 8 9 10
(2) 10 9 8 7 6 5 4 3 2 1

3 (continued)

3
(1) 2 — 4 — 6 — 8 — 10
(2) 2 — 4 — 6 — 8 — 10
(3) 2 — 4 — 6 — 8 — 10

4
(1) 1 2 3 4 5 6 7 8 9 10
(2) 1 2 3 4 5 6 7 8 9 10
(3) 1 2 3 4 5 6 7 8 9 10
(4) 2 4 6 8 10   (5) 3 2 1 0

4 Numbers up to 10 — pp 8,9

1 (1) ⓐ (2) ⓑ
(3) ⓑ (4) ⓐ

2 (1) ⓑ (2) ⓐ

3 (1) 4 (2) 7
(3) 8 (4) 10

4 (1) 3 (2) 2 (3) 7 (4) 10

5 (1) 3 (2) 4 (3) 7 (4) 9

5 Numbers up to 10 — pp 10,11

1 (1)

(2)

2 (1)
(2)

3 (1) ●●●●●●○○ (2) ○○○○●○○

4 (1) Third (2) Second
(3) David´s (4) Cathy´s

© Kumon Publishing Co., Ltd.

6 Numbers up to 10
pp 12,13

1 (1) ● (2) ● (3) ● (4) ●

2 (1) ● (2) ● (3) ● (4) ●

3 (1) 1 (2) 2 (3) 3 (4) 4

4 (1) 2 (2) 4 (3) 1 (4) 3 (5) 3
(6) 1 (7) 4 (8) 2

5 (1) 3 (2) 2 (3) 2 (4) 3 (5) 5
(6) 1 (7) 4 (8) 4 (9) 1 (10) 5

7 Numbers up to 10
pp 14,15

1 (1) ● (2) ● (3) ● (4) ● (5) ●

2 (1) ● (2) ● (3) ● (4) ● (5) ●

3 (1) 1 (2) 2 (3) 3 (4) 4 (5) 5

4 (1) 2 (2) 3 (3) 5 (4) 1 (5) 4
(6) 3 (7) 1 (8) 5 (9) 4 (10) 2

5 (1) 5 (2) 4 (3) 3 (4) 1 (5) 6

8 Numbers up to 10
pp 16,17

1 (1) ● (2) ● (3) ● (4) ● (5) ● (6) ●

2 (1) ● (2) ● (3) ● (4) ● (5) ● (6) ●

3 (1) 1 (2) 2 (3) 3 (4) 4 (5) 5 (6) 6

---

4 (1) 1 (2) 3 (3) 6 (4) 4 (5) 5 (6) 2
(7) 5 (8) 3 (9) 2 (10) 6 (11) 4 (12) 1

5 (1) 2 (2) 4 (3) 4 (4) 7

9 Numbers up to 10
pp 18,19

1 (1) ● (2) ● (3) ● (4) ● (5) ● (6) ● (7) ●

2 (1) ● (2) ● (3) ● (4) ● (5) ●

3 (1) 7 (2) 6 (3) 5 (4) 4 (5) 3 (6) 2

4 (1) 6 (2) 5 (3) 4 (4) 2 (5) 1
(6) 7 (7) 3 (8) 6 (9) 5 (10) 1
(11) 7 (12) 3 (13) 6 (14) 2

5 (1) 2 (2) 5 (3) 8 (4) 8

10 Numbers up to 10
pp 20,21

1 (1) ● (2) ● (3) ● (4) ● (5) ● (6) ● (7) ● (8) ●

2 (1) ● (2) ● (3) ● (4) ● (5) ● (6) ● (7) ●

3 (1) 8 (2) 6 (3) 5 (4) 3 (5) 2 (6) 1

4 (1) 7 (2) 4 (3) 1 (4) 5 (5) 8 (6) 6
(7) 3 (8) 2 (9) 6 (10) 5 (11) 1 (12) 3

5 (1) 7 (2) 4 (3) 7 (4) 9

© Kumon Publishing Co., Ltd.

11 Numbers up to 10
pp 22,23

1 (1) [dots] (2) [dots]
(3) [dots] (4) [dots]
(5) [dots] (6) [dots]
(7) [dots] (8) [dots]
(9) [dots]

2 (1) [dots] (2) [dots]
(3) [dots] (4) [dots]
(5) [dots]

3 (1) 9 (2) 8 (3) 7 (4) 6 (5) 5
(6) 4 (7) 3 (8) 1

4 (1) 7 (2) 5 (3) 2 (4) 6 (5) 3
(6) 9 (7) 8 (8) 4 (9) 1 (10) 7

5 (1) 8 (2) 3 (3) 10 (4) 10

12 Numbers up to 20
pp 24,25

1

11	12	13	14	15

16	17	18	19	20

15	13	17	16	11

19	12	20	13	18

2 (1) 12 (2) 13 (3) 14 (4) 18 (5) 11
(6) 15 (7) 12 (8) 19

13 Numbers up to 20
pp 26,27

1 (1) 1 (2) 2 (3) 3 (4) 4 (5) 5
(6) 6 (7) 7 (8) 8 (9) 9 (10) 10

2 (1) 10 (2) 10 (3) 10 (4) 10 (5) 10
(6) 10 (7) 10 (8) 10 (9) 10 (10) 10

3 (1) 11 (2) 12 (3) 13 (4) 14 (5) 15
(6) 16 (7) 17 (8) 18 (9) 19 (10) 20

4 (1) 5 (2) 7 (3) 8 (4) 4 (5) 10

5 (1) 16 (2) 12 (3) 19 (4) 11 (5) 13

14 Numbers up to 20
pp 28,29

1 (1)

1	2	3	4	5	6	7	8	9	10
11	12	13	14	15	16	17	18	19	20

(2)

1	2	3	4	5	6	7	8	9	10
11	12	13	14	15	16	17	18	19	20

(3)

1	2	3	4	5	6	7	8	9	10
11	12	13	14	15	16	17	18	19	20

(4)

1	2	3	4	5	6	7	8	9	10
11	12	13	14	15	16	17	18	19	20

(5)

1	2	3	4	5	6	7	8	9	10
11	12	13	14	15	16	17	18	19	20

2 (1) 10—11—12—13—14—15
(2) 15—16—17—18—19—20
(3) 20—19—18—17—16—15
(4) 14—13—12—11—10—9

3 (1) 8 9 10 11 12 13 14 15 16 17
(2) 20 19 18 17 16 15 14 13 12 11
(3) 18 17 16 15 14 13 12 11 10 9

4 (1) 12—13—14—15—16—17
(2) 18—17—16—15—14—13
(3) 10—12—14—16—18—20
(4) 20—18—16—14—12—10

15 Numbers up to 20
pp 30,31

1 (1) (From the left) 2, 6
(2) 0, 3, 8, 10
(3) 1, 7, 11, 14
(4) 3, 7, 10, 13, 19
(5) 0, 4, 8, 12, 18

 © Kumon Publishing Co., Ltd.

2 (1) (From the left) 5, 7, 12, 15
(2) 9, 13, 16, 18
(3) 2, 4, 6, 8, 10
(4) 2, 4, 6, 8, 10
(5) 0, 2, 4, 6, 8, 10, 12, 14, 16, 18, 20
(6) 0, 2, 4, 6, 8, 10, 12, 14, 16, 18, 20

16 Numbers up to 20
pp 32, 33

1 (1) 11 (2) 19 (3) 13 (4) 16 (5) 17 (6) 20
2 (1) 10 (2) 16 (3) 14 (4) 13 (5) 11 (6) 19
3 (1) 12 (2) 8 (3) 17 (4) 13
4 (1) 13 (2) 7 (3) 18 (4) 12

17 Numbers up to 120
pp 34, 35

1 (1) 16 (2) 19 (3) 22 (4) 27 (5) 34
(6) 38 (7) 42 (8) 45 (9) 53 (10) 59
(11) 64 (12) 66
2 (1) 43 (2) 34 (3) 78 (4) 57 (5) 86

18 Numbers up to 120
pp 36, 37

1 (1) 25 (2) 37 (3) 57 (4) 49 (5) 2, 8
(6) 8, 6 (7) 50 (8) 80
2 (1) 2 (2) 5
3 (1)
(2)

1	2	3	4	5	6	7	8	9	10
11	12	13	14	15	16	17	18	19	20
21	22	23	24	25	26	27	28	29	30
31	32	33	34	35	36	37	38	39	40
41	42	43	44	45	46	47	48	49	50
51	52	53	54	55	56	57	58	59	60
61	62	63	64	65	66	67	68	69	70
71	72	73	74	75	76	77	78	79	80
81	82	83	84	85	86	87	88	89	90
91	92	93	94	95	96	97	98	99	100
101	102	103	104	105	106	107	108	109	110
111	112	113	114	115	116	117	118	119	120

(3) 10, 20, 30, 40, 50, 60, 70, 80, 90, 100, 110, 120

19 Numbers up to 120
pp 38, 39

1 (1) (From the left) 1, 11, 16, 21, 26
(2) 22, 27, 32, 42, 48
(3) 25, 34, 37, 41, 45
(4) 41, 51, 61, 66, 76
(5) 45, 56, 62, 72, 78
(6) 64, 74, 84, 89, 94
2 (1) 12 (2) 8 (3) 22 (4) 18
3 (1) 28 (2) 39 (3) 70 (4) 91
4 (1) 32 (2) 58 (3) 80 (4) 99

20 Numbers up to 120
pp 40, 41

1 (1) ⓑ
2 (1) 21 (2) 27
(3) 33 (4) 46
(5) 48 (6) 66
3 (1) 51 (2) 82
(3) 83 (4) 64
(5) 100 (6) 98
4 (1) 23, 21, 12 (2) 90, 19, 9
(3) 74, 71, 47 (4) 96, 89, 69

21 Telling Time
pp 42, 43

1 (1) 1:00 (2) 2:00 (3) 3:00 (4) 5:00
(5) 4:00 (6) 6:00
2 (1) 8:00 (2) 10:00 (3) 11:00 (4) 12:00
3 (1) 9:30 (2) 11:30
(3) 1:30 (4) 5:30
(5) 3:30 (6) 12:30
4 (1) 2:30 (2) 4:30
(3) 8:30 (4) 10:30

22 Length pp 44,45

1 ⓐ ✕ ⓑ ✓ ⓒ ✕ ⓓ ✕ ⓔ ✕

2 ⓐ 2 ⓑ 1 ⓒ 3

3 (1) ⓐ (2) ⓑ
  (3) ⓐ (4) ⓑ
  (5) ⓐ

4 ⓐ 3 ⓑ 1 ⓒ 2

23 Length pp 46,47

1 (1) ⓐ (2) ⓑ

2 (1) ⓑ (2) ⓑ
  (3) ⓑ (4) ⓑ

3 (1) Chopsticks
  (2) Soap

4 (1) green, 2
  (2) green, 2

24 Weight pp 48,49

1 (1)

 (2)

 (3) (4)

2 (1) B (2) A
  (3) A (4) B

3 (1) 3 (2) 5 (3) 7
  (4) 8 (5) 12

25 Area pp 50,51

1 (1) B (2) A
  (3) A (4) B

2 (1) D (2) C

3 (1) 2 (2) 4 (3) 5 (4) 5
  (5) 5 (6) 8 (7) 5 (8) 8

26 Volume pp 52,53

1 (1) A (2) B (3) B (4) A

2 (1) A (2) A (3) B (4) B

3 (1) 3 (2) 6 (3) 2 (4) 8
  (5) 4 (6) 5

27 Temperature pp 54,55

1 (1) A (2) B (3) B (4) A

2 (1) B (2) B (3) A (4) A

3 (1) The warmest : D   The coldest : A
  (2) The warmest : B   The coldest : C
  (3) The warmest : A   The coldest : E

28 Coins pp 56,57

1 (1) 1 (2) 1 (3) 2 (4) 2 (5) 2

2 (1) 1 (2) 2 (3) 3 (4) 4 (5) 5
  (6) 6 (7) 7 (8) 8 (9) 9 (10) 10

3 (1) 5 (2) 5 (3) 10 (4) 10 (5) 10

4 (1) 5 (2) 10 (3) 15 (4) 20 (5) 25
  (6) 30 (7) 35 (8) 40 (9) 45 (10) 50

29 Coins pp 58,59

1 (1) 10 (2) 10 (3) 20 (4) 20 (5) 20

2 (1) 10 (2) 20 (3) 30 (4) 40 (5) 50
  (6) 60 (7) 70 (8) 80 (9) 90 (10) 100

3 (1) 25 (2) 25 (3) 50 (4) 50 (5) 50

4 (1) 25 (2) 50 (3) 75 (4) 100 (5) 125
  (6) 150 (7) 175 (8) 200 (9) 225 (10) 250

30 Coins pp 60,61

1 (1) 3 (2) 5 (3) 15 (4) 5 (5) 20
  (6) 25 (7) 25 (8) 40 (9) 20 (10) 75

2 (1) 7 (2) 21 (3) 45 (4) 35 (5) 53
  (6) 35 (7) 22 (8) 50 (9) 32 (10) 41

© Kumon Publishing Co., Ltd.

31 Shapes
pp 62,63

1 ⓒ

2 (1) ⓑ, ⓓ   (2) ⓐ, ⓒ
(3) ⓑ   (4) ⓓ

3
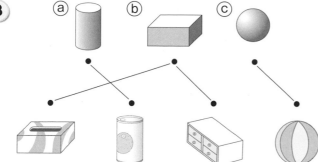

4 (1) 3   (2) 2   (3) 1   (4) 3

32 Shapes
pp 64,65

1 (1) ⓒ   (2) ⓑ
(3) ⓓ

2
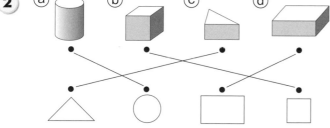

3 ⓐ, ⓑ

4 (1) ⓑ, ⓓ   (2) ⓒ, ⓓ

5
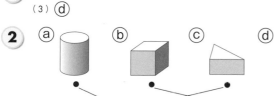

33 Shapes
pp 66,67

1 (1) 2 pieces   (2) 3 pieces   (3) 4 pieces

2
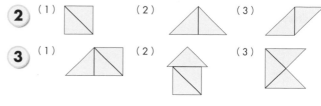

3

4 (1) 6 pieces   (2) 5 pieces

5 (1) 4 pieces   (2) 10 pieces   (3) 9 pieces

34 Shapes
pp 68,69

1 (1) ⓒ   (2) ⓒ

2 (1) ⓐ   (2) ⓓ

3 ⓑ

4
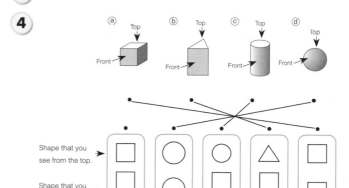

35 Review
pp 70,71

1 (1) 67   (2) 54

2 (1) 76   (2) 100   (3) 6, 8   (4) 8
(5) 4, 9   (6) 30

3 (1) 52   (2) 86

4 (1) ⓑ   (2) ⓑ

5
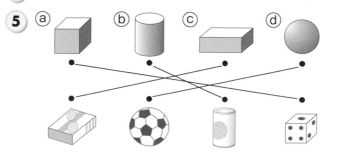

© Kumon Publishing Co., Ltd.   79

1 (1) (From the left) 0, 6, 9, 14, 19

(2) 43, 57, 79, 88, 96

2 (1) 43, 34, 33   (2) 97, 89, 79

3 (1) 62   (2) 58   (3) 41   (4) 37

4 (1) | 25 | 26 | 27 | 28 | 29 | 30 | 31 | 32 |

(2) | 78 | 79 | 80 | 81 | 82 | 83 | 84 | 85 |

(3) | 100 | 99 | 98 | 97 | 96 | 95 | 94 | 93 |

(4) | 64 | 66 | 68 | 70 | 72 | 74 | 76 | 78 |

(5) | 23 | 33 | 43 | 53 | 63 | 73 | 83 | 93 |

5 (1) ⓐ, ⓒ

(2) ⓑ, ⓒ

 © Kumon Publishing Co., Ltd.